DYNAMICAL THEORIES OF BROWNIAN MOTION

BY

EDWARD NELSON

PRINCETON, NEW JERSEY
PRINCETON UNIVERSITY PRESS
1967

Printed in the United States of America

CONTENTS*

*These are the notes for the second term of a year course on stochastic processes. The audience was familiar with Markoff processes, martingales, the detailed nature of the sample paths of the Wiener process, and measure theory on the space of sample paths.

After some historical and elementary material in §§1 - 8, we discuss the Ornstein-Uhlenbeck theory of Brownian motion in §§9 - 10, showing that the Einstein-Smoluchowski theory is in a rigorous and strong sense the limiting theory for infinite friction. The results of the long §11 are not used in the following, except for the concepts of mean forward velocity (p.80), mean backward velocity (p.95), and mean acceleration (p.99). The rest of the notes deal with probability theory in quantum mechanics and in the alternative stochastic theory due to Imre Fényes and others.

I wish to express my thanks to Lionel Rebhun for showing us the Brownian motion of a colloidal particle, to many members of the class for their lively and critical participation, to Elizabeth Epstein for a beautiful job of typing, and to the National Science Foundation (Grant GP5264) for support during part of the time when these notes were written.

§1. Apology

It is customary in Fine Hall to lecture on mathematics, and any major deviation from that custom requires a defense.

It is my intention in these lectures to focus on Brownian motion as a natural phenomenon. I will review the theories put forward to account for it by Einstein, Smoluchowski, Langevin, Ornstein, Uhlenbeck, and others. It will be my conjecture that a certain portion of current physical theory, while mathematically consistent, is physically wrong, and I will propose an alternative theory.

Clearly, the chances of this conjecture being correct are exceedingly small, and since the contention is not a mathematical one, what is the justification for spending time on it? The presence of some physicists in the audience is irrelevant. Physicists lost interest in the phenomenon of Brownian motion about thirty or forty years ago. If a modern physicist is interested in Brownian motion, it is because the mathematical theory of Brownian motion has proved useful as a tool in the study of some models of quantum field theory and in quantum statistical mechanics. I believe that this approach has exciting possibilities, but I will not deal with it in this course (though some of the mathematical techniques which will be developed are relevant to these problems).

The only legitimate justification is a mathematical one. Now "applied mathematics" contributes nothing to mathematics. On the other hand, the sciences and technology do make vital contributions to mathematics. The ideas in analysis which had their origin in physics are so numerous and so central that analysis would be unrecognizable without them.

A few years ago topology was in the doldrums, and then it was revitalized by the introduction of differential structures. A significant role in this process is being played by the qualitative theory of ordinary differential equations, a subject having its roots in science and technology. There was opposition on the part of some topologists to this process, due to the loss of generality and the impurity of methods.

It seems to me that the theory of stochastic processes is in the doldrums today. It is in the doldrums for the same reason, and the remedy is the same. We need to introduce differential structures and accept the corresponding loss of generality and impurity of methods. I hope that a study of dynamical theories of Brownian motion can help in this process.

Professor Rebhun has very kindly prepared a demonstration of Brownian motion in Moffet Laboratory. This is a live telecast from a microscope. It consists of carmine particles in acetone, which has lower viscosity than water. The smaller particles have a diameter of about two microns (a micron is one thousandth of a millimeter). Notice that they are more active than the larger particles. The other sample consists of carmine particles in water - they are considerably less active. According to theory, nearby particles are supposed to move independently of each other, and this appears to be the case.

§1. APOLOGY

Perhaps the most striking aspect of actual Brownian motion is the apparent tendency of the particles to dance about without going anywhere. Does this accord with theory, and how can it be formulated?

One nineteenth century worker in the field wrote that although the terms "titubation" and "pedesis" were in use, he preferred "Brownian movements" since everyone at once knew what was meant. (I looked up these words [1]. Titubation is defined as the "act of titubating; specif., a peculiar staggering gait observed in cerebellar and other nervous disturbances." The definition of pedesis reads, in its entirety, "Brownian movement.") Unfortunately, this is no longer true, and semantical confusion can result. I shall use "Brownian motion" to mean the natural phenomenon. The common mathematical model of it will be called (with ample historical justification) the "Wiener process."

I plan to waste your time by considering the history of nineteenth century work on Brownian motion in unnecessary detail. We will pick up a few facts worth remembering when the mathematical theories are discussed later, but only a few. Studying the development of a topic in science can be instructive. One realizes what an essentially comic activity scientific investigation is (good as well as bad).

Reference

[1]. Webster's New International Dictionary, Second Edition, G. & C. Merriam Co., Springfield, Mass. (1961).

§2. Robert Brown

Robert Brown sailed in 1801 to study the plant life of the coasts of Australia. This was only a few years after a botanical expedition to Tahiti aboard the Bounty ran into unexpected difficulties. Brown returned to England in 1805, however, and became a distinguished botanist. Although Brown is remembered by mathematicians only as the discoverer of Brownian motion, his biography in the Encyclopaedia Britannica makes no mention of this discovery.

Brown did not discover Brownian motion. After all, practically anyone looking at water through a miscroscope is apt to see little things moving around. Brown himself mentions one precursor in his 1828 paper [2] and ten more in his 1829 paper [3], starting at the beginning with Leeuwenhoek (1632-1723), including Buffon and Spallanzani (the two protagonists in the eighteenth century debate on spontaneous generation), and one man (Bywater, who published in 1819) who reached the conclusion (in Brown's words) that "not only organic tissues, but also inorganic substances, consist of what he calls animated or irritable particles."

The first dynamical theory of Brownian motion was that the particles were alive. The problem was in part observational, to decide whether a particle is an organism, but the vitalist bugaboo was mixed up in it. Writing as late as 1917, D'Arcy Thompson [4] observes: "We cannot, indeed, without the most careful scrutiny, decide whether the movements of our minutest organisms are intrinsically 'vital' (in the sense of being beyond a physical mechanism, or working model) or not." Thompson describes some motions of minute organisms, which had been ascribed to their own activity, but which he says can be explained in terms of the physical picture of Brownian motion as due to molecular bombardment. On the other hand, Thompson describes an experiment by Karl Przibram, who observed the position of a unicellular organism at fixed intervals. The organism was much too active, for a body of its size, for its motion to be attributed to molecular bombardment, but Przibram concluded that, with a suitable choice of diffusion coefficient, Einstein's law applied!

Although vitalism is dead, Brownian motion continues to be of interest to biologists. Some of you heard Professor Rebhun describe the problem of disentangling the Brownian component of some unexplained particle motions in living cells.

Some credit Brown with showing that the Brownian motion is not vital in origin; others appear to dismiss him as a vitalist. It is of interest to follow Brown's own account [2] of his work. It is one of those rare papers in which a scientist gives a lucid step-by-step account of his discovery and reasoning.

Brown was studying the fertilization process in a species of flower which, I believe likely, was discovered on the Lewis and Clark expedition. Looking at the pollen in water through a microscope, he observed small particles in "rapid oscillatory motion." He then examined pollen of other species, with similar results. His first hypothesis was that Brownian motion was not only vital but peculiar to the male sexual cells of plants. (This we know is not true - the carmine particles which we saw were derived from the dried bodies of female insects which grow on cactus plants in Mexico and Central America.) Brown describes how this view was modified:

"In this stage of the investigation having found, as I believed, a peculiar character in the motions of the particles of pollen in water, it occurred to me to appeal to this peculiarity as a test in certain Cryptogamous plants, namely Mosses, and the genus *Equisetum*, in which the existence of sexual organs had not been universally admitted. ... But I at the same time observed, that on bruising the ovula or seeds of *Equisetum*, which at first happened accidentally, I so greatly increased the number of moving particles, that the source of the added quantity could not be doubted. I found also that on bruising first the floral leaves of Mosses, and then all other parts of those plants, that I readily obtained similar particles, not in equal quantity indeed, but equally in motion. My supposed test of the male organ was therefore necessarily abandoned.

"Reflecting on all the facts with which I had now become acquainted, I was disposed to believe that the minute spherical particles or Molecules of apparently uniform size,... were in reality the supposed constituent or elementary molecules of organic bodies, first so considered by Buffon and Needham ..."

He examined many organic substances, finding the motion, and then looked at mineralized vegetable remains: "With this view a minute portion of silicified wood, which exhibited the structure of Coniferae, was bruised, and spherical particles, or molecules in all respects like those so frequently mentioned, were readily obtained from it; in such quantity, however, that the whole substance of the petrifaction seemed to be formed of them. From hence I inferred that these molecules were not limited to organic bodies, nor even to their products."

He tested this inference on glass and minerals: "Rocks of all ages, including those in which organic remains have never been found, yielded the molecules in abundance. Their existence was ascertained in each of the constituent minerals of granite, a fragment of the Sphinx being one of the specimens observed."

Brown's work aroused widespread interest. We quote from a report [5] published in 1830 of work of Muncke in Heidelberg:

"This motion certainly bears some resemblance to that observed in infusory animals, but the latter show more of a voluntary action. The idea of vitality is quite out of the question. On the contrary, the motions may be viewed as of a mechanical nature, caused by the unequal temperature of the strongly illuminated water, its evaporation, currents of air, and heated currents, &c."

Of the causes of Brownian motion, Brown [3] writes:

"I have formerly stated my belief that these motions of the particles neither arose from currents in the fluid containing them, nor depended on that intestine motion which may be supposed to accompany its evaporation.

"These causes of motion, however, either singly or combined with other, - as, the attractions and repulsions among the particles themselves, their unstable equilibrium in the fluid in which they are suspended, their hygrometrical or capillary action, and in some cases the disengagement of volatile matter, or of minute air bubbles, - have been considered by several writers as sufficiently accounting for the appearances."

He refutes most of these explanations by describing an experiment in which a drop of water of microscopic size immersed in oil, and containing as few as one particle, exhibits the motion unabated.

Brown denies having stated that the particles are animated. His theory, which he is careful never to state as a conclusion, is that matter is composed of small particles, which he calls active molecules, which exhibit a rapid, irregular motion having its origin in the particles themselves and not in the surrounding fluid.

His contribution was to establish Brownian motion as an important phenomenon, to demonstrate clearly its presence in inorganic as well as organic matter, and to refute by experiment facile mechanical explanations of the phenomenon.

References

[2]. Robert Brown, A brief Account of Microscopical Observations made in the Months of June, July, and August, 1827, on the Particles contained in the Pollen of Plants; and on the general Existence of active Molecules in Organic and Inorganic Bodies, Philosophical Magazine N.S. 4(1828), 161-173.

[3]. Robert Brown, Additional Remarks on Active Molecules, Philosophical Magazine N.S. 6(1829), 161-166.

[4]. D'Arcy W. Thompson, Growth and Form, Cambridge University Press (1917).

[5]. Intelligence and Miscellaneous Articles: Brown's Microscopical Observations on the Particles of Bodies, Philosophical Magazine N.S. 8(1830), 296.

§3. The period before Einstein

I have found no reference to a publication on Brownian motion between 1831 and 1857. Reading papers published in the sixties and seventies, however, one has the feeling that awareness of the phenomenon remained widespread (it could hardly have failed to, as it was something of a nuisance to microscopists). Knowledge of Brown's work reached literary circles. In George Eliot's "Middlemarch" (Book II, Chapter V, published in 1872) a visitor to the vicar is interested in obtaining one of the vicar's biological specimens and proposes a barter: "I have some sea mice. ... And I will throw in Robert Brown's new thing, - 'Microscopic Observations on the Pollen of Plants,' - if you don't happen to have it already."

From the eighteen sixties on, many scientists worked on the phenomenon. Most of the hypotheses which were advanced could have been ruled out by consideration of Brown's experiment on the microscopic water drop enclosed in oil. The first to express a notion close to the modern theory of Brownian motion was Wiener in 1863. A little later Carbonelle claimed that the internal movements which constitute the heat content of fluids is well able to account for the facts. A passage emphasizing the probabilistic aspects is quoted by Perrin [6, p.4]:

"In the case of a surface having a certain area, the molecular collisions of the liquid which cause the pressure, would not produce any perturbation of the suspended particles, because these, as a whole, urge the particles equally in all directions. But if the surface is of area less than is necessary to ensure the compensation of irregularities, there is no longer any ground for considering the mean pressure; the inequal pressures, continually varying from place to place, must be recognized, as the law of large numbers no longer leads to uniformity; and the resultant will not now be zero but will change continually in intensity and direction. Further, the inequalities will become more and more apparent the smaller the body is supposed to be, and in consequence the oscillations will at the same time become more and more brisk ..."

There was no unanimity in this view. Jevons maintained that pedesis was electrical in origin. Ord, who attributed Brownian motion largely to "the intestine vibration of colloids," attacked Jevons' views [7], and I cannot refrain from quoting him:

"I may say that before the publication of Dr. Jevons' observations I had made many experiments to test the influence of acids [upon Brownian movements], and that my conclusions entirely agree with his. In stating this, I have no intention of derogating from the originality of Professor Jevons, but simply of adding my testimony to his on a matter of some importance. ...

"The influence of solutions of soap upon Brownian movements, as set forth by Professor Jevons, appears to me to support my contention in the way of agreement. He shows that the introduction of soap in the suspending fluid quickens and makes more persistent the movements of the suspended particles. Soap in the eyes of Professor Jevons acts conservatively by retaining or not conducting electricity. In my eyes it is a colloid, keeping up movements by revolutionary perturbations. ... It is interesting to remember that, while soap is probably our best detergent, boiled oatmeal is one of its best substitutes. What this may be as a conductor of electricity I do not know, but it certainly is a colloid mixture or solution."

Careful experiments and arguments supporting the kinetic theory were made by Gouy. From his work and the work of others emerged the following main points (cf. [6]):

1) The motion is very irregular, composed of translations and rotations, and the trajectory appears to have no tangent.

2) Two particles appear to move independently, even when they approach one another to within a distance less than their diameter.

3) The motion is more active the smaller the particles.

4) The composition and density of the particles have no effect.

5) The motion is more active the less viscous the fluid.

6) The motion is more active the higher the temperature.

7) The motion never ceases.

In discussing 1), Perrin mentions the mathematical existence of nowhere differentiable curves. Point 2) had been noticed by Brown, and it is a strong argument against gross mechanical explanations. Perrin points out that 6) (although true) had not really been established by observation, since for a given fluid the viscosity usually increases by a greater factor than the absolute temperature, so that the effect 5) dominates 6). Point 7) was established by observing a sample over a period of twenty years, and by observations of liquid inclusions in quartz thousands of years old. This point rules out all attempts to explain Brownian motion as a non-equilibrium phenomenon.

By 1905, the kinetic theory, that Brownian motion of microscopic particles is caused by bombardment by the molecules of the fluid, seemed the most plausible. The seven points mentioned above did not seem to be in conflict with this theory. The kinetic theory appeared to be open to a simple test: the law of equipartition of energy in statistical mechanics implied that the kinetic energy of translation of a particle and of a molecule should be equal. The latter was roughly known (by a determination of Avogadro's number by other means), the mass of a particle could be determined, so all one had to measure was the velocity of a particle in Brownian motion. This was attempted by several experimenters, but the result failed to confirm the kinetic theory as the two values of kinetic energy differed by a factor of about 100,000. The difficulty, of course, was point 1) above. What is meant by the velocity of a Brownian particle? This is a question which will recur throughout these lectures. The success of Einstein's theory of Brownian motion (1905) was largely due to his circumventing this question.

References

[6]. Jean Perrin, Brownian movement and molecular reality, translated from the Annales de Chimie et de Physique, 8me Series, 1909,

by F. Soddy, Taylor and Francis, London, 1910.

[7]. William M. Ord, M.D., On some Causes of Brownian Movements, Journal of the Royal Microscopical Society, 2(1879), 656-662.

The following also contain historical remarks (in addition to [6]). You are advised to consult at most one account, since they contradict each other not only in interpretation but in the spelling of the names of some of the men involved.

[8]. Jean Perrin, Atoms, translated by D. A. Hammick, Van Nostrand, 1916. (Chapters III and IV deal with Brownian motion, and they are summarized in the author's article Brownian Movement in the Encyclopaedia Britannica.)

[9]. E. F. Burton, The Physical Properties of Colloidal Solutions, Longmans, Green and Co., London, 1916. (Chapter IV is entitled The Brownian Movement. Some of the physics in this chapter is questionable.)

[10]. Albert Einstein, Investigations on the Theory of the Brownian Movement, edited with notes by R. Fürth, translated by A. D. Cowper, Dover, 1956. (Fürth's first note, pp. 86-88, is historical.)

[11]. R. Bowling Barnes and S. Silverman, Brownian Motion as a Natural Limit to all Measuring Processes, Reviews of Modern Physics 6(1934), 162-192.

§4. Albert Einstein

It is sad to realize that despite all of the hard work which had gone into the study of Brownian motion, Einstein was unaware of the existence of the phenomenon. He predicted it on theoretical grounds and formulated a correct quantitative theory of it. (This was in 1905, the same year he discovered the special theory of relativity and invented the photon.) As he describes it [12, p.47]:

"Not acquainted with the earlier investigations of Boltzmann and Gibbs, which had appeared earlier and actually exhausted the subject, I developed the statistical mechanics and the molecular-kinetic theory of thermodynamics which was based on the former. My major aim in this was to find facts which would guarantee as much as possible the existence of atoms of definite finite size. In the midst of this I discovered that, according to atomistic theory, there would have to be a movement of suspended microscopic particles open to observation, without knowing that observations concerning the Brownian motion were already long familiar."

By the time his first paper on the subject was written, he had heard of Brownian motion [10, §3, p.1]:

"It is possible that the movements to be discussed here are identical with the so-called 'Brownian molecular motion'; however, the information available to me regarding the latter is so lacking in precision, that I can form no judgment in the matter."

There are two parts to Einstein's argument. The first is mathematical and will be discussed later (§5). The result is the following: Let $\rho=\rho(x,t)$ be the probability density that a Brownian particle is at x at time t. Then, making certain probabilistic assumptions (some of them implicit), Einstein derived the diffusion equation

$$\frac{\partial\rho}{\partial t} = D\Delta\rho \tag{1}$$

where D is a positive constant, called the coefficient of diffusion. If the particle is at 0 at time 0 (so that $\rho(x,0)=\delta(x)$) then

$$\rho(x,t) = \frac{1}{(4\pi Dt)^{3/2}} e^{-\frac{|x|^2}{4Dt}} \tag{2}$$

(in three dimensional space, where $|x|$ is the Euclidean distance of x from the origin).

The second part of the argument, which relates D to other physical quantities, is physical. In essence, it runs as follows. Imagine a suspension of many Brownian particles in a fluid, acted on by an external force K, and in equilibrium. (The force K might be gravity, as in the figure, but the beauty of the argument is that K is entirely virtual.)

In equilibrium, the force K is balanced by the osmotic pressure forces of the suspension,

$$K = kT \frac{\operatorname{grad} \nu}{\nu} . \tag{3}$$

Here ν is the number of particles per unit volume, T is the absolute temperature, and k is Boltzmann's constant. Boltzmann's constant has the dimensions of energy per degree, so that kT has

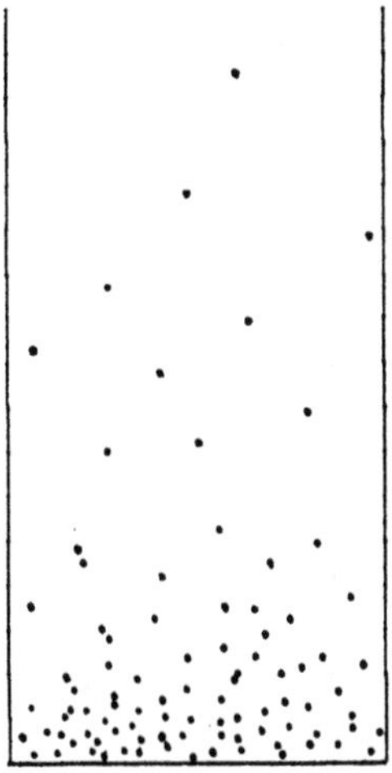

Figure 1

the dimensions of energy. A knowledge of k is equivalent to a knowledge of Avogadro's number, and hence of molecular sizes. The right hand side of (3) is derived by applying to the Brownian particles the same considerations that are applied to gas molecules in the kinetic theory.

The Brownian particles moving in the fluid experience a resistance due to friction, and the force K imparts to each particle a velocity of the form

$$\frac{K}{m\beta},$$

where β is a constant with the dimensions of frequency (inverse time) and m is the mass of the particle. Therefore

$$\frac{\nu K}{m\beta}$$

particles pass a unit area per unit of time due to the action of the force K. On the other hand, if diffusion alone were acting, ν would satisfy the diffusion equation

$$\frac{\partial \nu}{\partial t} = D\Delta\nu$$

so that

$$-D \text{ grad } \nu$$

particles pass a unit area per unit of time due to diffusion. In dynamical equilibrium, therefore,

$$\frac{\nu K}{m\beta} = D \text{ grad } \nu \ . \tag{4}$$

Now we may eliminate K and ν between (3) and (4), giving Einstein's formula

$$D = \frac{kT}{m\beta} \ . \tag{5}$$

This formula applies even when there is no force and when there is only one Brownian particle (so that ν is not defined).

Parenthetically, if we divide both sides of (3) by $m\beta$, and use (5), we obtain

$$\frac{K}{m\beta} = D \frac{\text{grad } \nu}{\nu} \ .$$

The probability density ρ is just the number density ν divided by the total number of particles, so this may be rewritten as

$$\frac{K}{m\beta} = D \frac{\text{grad } \rho}{\rho} \ .$$

Since the left hand side is the velocity acquired by a particle due to the action of the force,

$$D \frac{\text{grad } \rho}{\rho} \tag{6}$$

is the velocity required of the particle to counteract osmotic effects.

If the Brownian particles are spheres of radius a, then Stokes' theory of friction gives $m\beta = 6\pi\eta a$, where η is the coefficient of viscosity of the fluid, so that in this case

$$D = \frac{kT}{6\pi\eta a}. \tag{7}$$

The temperature T and the coefficient of viscosity η may be measured, with great labor a colloidal suspension of spherical particles of fairly uniform radius a may be prepared, and D may be determined by statistical observations of Brownian motion using (2). In this way Boltzmann's constant k (or, equivalently Avogadro's number) can be determined. This was done in a series of difficult and laborious experiments by Perrin and Chaudesaigues [6, §3]. Rather surprisingly, considering the number of assumptions which went into the argument, the result obtained for Avogadro's number agreed to within 19% of the modern value obtained by other means. Notice how the points 3) - 6) of §3 are reflected in the formula (7).

Einstein's argument does not give a dynamical theory of Brownian motion; it only determines the nature of the motion and the value of the diffusion coefficient on the basis of some assumptions. Smoluchowski, independently of Einstein, attempted a dynamical theory, and arrived at (5) with a factor of 32/27 on the right hand side. Langevin gave another derivation of (5) which was the starting point for the work of Ornstein and Uhlenbeck, which we shall discuss later (§§9-10). Langevin is the founder of the theory of stochastic differential equations (which is the subject matter of these lectures).

Einstein's work was of great importance in physics, for it showed in a visible and concrete way that atoms are real. Quoting from Einstein's Autobiographical Notes again [12, p.49]:

"The agreement of these considerations with experience together with Planck's determination of the true molecular size from the law of radiation (for high temperatures) convinced the sceptics, who were quite numerous at that time (Ostwald, Mach) of the reality of atoms. The antipathy of these scholars towards atomic theory can indubitably be traced back to their positivistic philosophical attitude. This is an interesting example of the fact that even scholars of audacious spirit and fine instinct can be obstructed in the interpretation of facts by philosophical prejudices."

Let us not be too hasty in adducing any other interesting example which may spring to mind.

Reference

[12]. Paul Arthur Schilpp, editor, Albert Einstein: Philosopher-Scientist, The Library of Living Philosophers, Vol. VII, The Library of Living Philosophers, Inc., Evanston, Illinois, 1949.

§5. Derivation of the Wiener process

Einstein's basic assumption is that the following is possible [10, §3, p.13]: "We will introduce a time-interval τ in our discussion, which is to be very small compared with the observed interval of time [i.e., the interval of time between observations], but, nevertheless, of such a magnitude that the movements executed by a particle in two consecutive intervals of time τ are to be considered as mutually independent phenomena." He then implicitly considers the limiting case $\tau \to 0$. This assumption has been criticized by many people, including Einstein himself, and later on (§§9-10) we will discuss a theory in which this assumption is modified. Einstein's derivation of the transition probabilities proceeds by formal manipulations of power series. His neglect of higher order terms is tantamount to the assumption (ii) below. In the theorem below, p^t may be thought of as the probability distribution at time t of the x-coordinate of a Brownian particle starting at x=0 at t=0. The proof is taken from a paper of Hunt [13], who showed that Fourier analysis is not the natural tool for problems of this type.

Theorem 5.5. *Let* p^t, $0 \le t < \infty$, *be a family of probability measures on the real line* $\mathbb{R}$ *such that*

$$\text{(i)} \quad p^t * p^s = p^{t+s}; \qquad 0 \le t,s < \infty,$$

where $*$ *denotes convolution; for each* $\varepsilon > 0$,

$$\text{(ii)} \quad p^t(\{x: |x| \ge \varepsilon\}) = o(t), \qquad t \to 0;$$

and for each $t > 0$, p^t *is invariant under the transformation* $x \to -x$. *Then either* $p^t = \delta$ *for all* $t \geq 0$ *or there is a* $D > 0$ *such that, for all* $t > 0$, p^t *has the density*

$$p(t,x) = \frac{1}{\sqrt{4\pi Dt}} e^{-\frac{x^2}{4Dt}},$$

so that p *satisfies the diffusion equation*

$$\frac{\partial p}{\partial t} = D \frac{\partial^2}{\partial x^2} p, \qquad t > 0.$$

First we need a lemma:

Theorem 5.1. Let $\mathfrak{X}$ *be a real Banach space,* $f \in \mathfrak{X}$, $\mathfrak{D}$ *a dense linear subspace of* $\mathfrak{X}$, $u_1, \ldots, u_n$ *continuous linear functionals on* $\mathfrak{X}$, $\delta > 0$. *Then there exists a* $g \in \mathfrak{D}$ *with*

$$\|f-g\| \leq \delta$$

$$(u_1,f) = (u_1,g), \ldots, (u_n,f) = (u_n,g).$$

Proof. Let us instead prove that if $\mathfrak{X}$ is a real Banach space, $\mathfrak{D}$ a dense convex subset, $\mathfrak{M}$ a closed affine hyperplane, then $\mathfrak{D} \cap \mathfrak{M}$ is dense in $\mathfrak{M}$. Then the general case of finite co-dimension follows by induction.

Without loss of generality, we may assume that $\mathfrak{M}$ is linear ($0 \in \mathfrak{M}$), so that, if we let e be an element of $\mathfrak{X}$ not in $\mathfrak{M}$,

$$\mathfrak{X} = \mathfrak{M} \oplus \mathbb{R}e.$$

Let $f \in \mathfrak{M}$, $\varepsilon > 0$. Choose g_+ in $\mathfrak{D}$ so that

$$\|(f+e) - g_+\| \leq \varepsilon$$

and choose g_- in $\mathcal{D}$ so that

$$\|(f-e) - g_-\| \leq \varepsilon.$$

Set

$$g_+ = m_+ + r_+ e, \quad m_+ \in \mathfrak{M}$$
$$g_- = m_- + r_- e, \quad m_- \in \mathfrak{M}$$

Since $\mathfrak{M}$ is closed, the linear functional which assigns to each element of $\mathfrak{X}$ the corresponding coefficient of e is continuous. Therefore r_+ and r_- tend to 1 as $\varepsilon \to 0$ and so are strictly positive for ε sufficiently small. By the convexity of $\mathcal{D}$,

$$g = \frac{r_- g_+ + r_+ g_-}{r_- + r_+}$$

is then in $\mathcal{D}$. But

$$g = \frac{r_- m_+ + r_+ m_-}{r_- + r_+}$$

is also in $\mathfrak{M}$, and it converges to f as $\varepsilon \to 0$. QED.

We recall that if $\mathfrak{X}$ is a Banach space, then a <u>contraction semigroup</u> on $\mathfrak{X}$ (in our terminology) is a family of bounded linear transformations P^t of $\mathfrak{X}$ into itself, defined for $0 \leq t < \infty$, such that $P^0=1$, $P^t P^s = P^{t+s}$, $\|P^t f - f\| \to 0$, and $\|P^t\| \leq 1$, for all $0 \leq t, s < \infty$ and all f in $\mathfrak{X}$. The <u>infinitesimal generator</u> A is defined by

$$Af = \lim_{t \to 0^+} \frac{P^t f - f}{t}$$

on the domain (A) of all f for which the limit exists.

If X is a locally compact Hausdorff space, $C(X)$ denotes the Banach space of all continuous functions vanishing at infinity in the norm

$$\|f\| = \sup_{x \in X} |f(x)|,$$

and $\dot{X}$ denotes the one-point compactification of X. We denote by $C^2_{com}(\mathbb{R}^\ell)$ the set of all functions of class C^2 with compact support on $\mathbb{R}^\ell$, by $C^2(\mathbb{R}^\ell)$ its completion in the norm

$$\|f\|' = \|f\| + \sum_{i=1}^{\ell} \left\|\frac{\partial f}{\partial x^i}\right\| + \sum_{i,j=1}^{\ell} \left\|\frac{\partial^2 f}{\partial x^i \partial x^j}\right\|,$$

and by $C^2(\dot{\mathbb{R}}^\ell)$ the completion of $C^2_{com}(\mathbb{R}^\ell)$ together with the constants, in the same norm.

A <u>Markovian semigroup</u> on $C(X)$ is a contraction semigroup on $C(X)$ such that $f \geq 0$ implies $P^t f \geq 0$ for $0 \leq t < \infty$, and such that for all x in X and $0 < t < \infty$,

$$\sup_{\substack{0 \leq f \leq 1 \\ f \in C(X)}} P^t f(x) = 1 .$$

If X is compact, the last condition is equivalent to $P^t 1 = 1$, $0 \leq t < \infty$. By the Riesz theorem, there is a unique regular Borel probability measure $p^t(x,\cdot)$ such that

$$P^t f(x) = \int p^t(x,dy) f(y),$$

and p^t is called the <u>kernel</u> of P^t.

<u>Theorem 5.2.</u> <u>Let</u> P^t <u>be a Markovian semigroup on</u> $C(\dot{\mathbb{R}}^\ell)$ <u>commuting with translations, and let</u> A <u>be the infinitesimal generator of</u> P^t. <u>Then</u>

$$C^2(\dot{\mathbb{R}}^\ell) \subset \mathcal{D}(A).$$

Proof. Since P^t commutes with translations, P^t leaves $C^2(\dot{\mathbb{R}}^\ell)$ invariant and is a contraction semigroup on it. Let A' be the infinitesimal generator of P^t on $C^2(\dot{\mathbb{R}}^\ell)$. Clearly $\mathcal{D}(A') \subset \mathcal{D}(A)$, and since the domain of the infinitesimal generator is always dense, $\mathcal{D}(A) \cap C^2(\dot{\mathbb{R}}^\ell)$ is dense in $C^2(\dot{\mathbb{R}}^\ell)$.

Let ψ be in $C^2(\dot{\mathbb{R}}^\ell)$ and such that $\psi(x)=|x|^2$ in a neighborhood of 0, $\psi(x)=1$ in a neighborhood of infinity, and ψ is strictly positive on $\dot{\mathbb{R}}^\ell-\{0\}$. Apply Theorem 5.1 to $\mathfrak{X} = C^2(\dot{\mathbb{R}}^\ell)$, $\mathcal{D} = \mathcal{D}(A) \cap C^2(\dot{\mathbb{R}}^\ell)$, $f=\psi$, and to the continuous linear functionals given by

$$\varphi(0), \quad \frac{\partial\varphi}{\partial x^i}(0), \quad \frac{\partial^2\varphi}{\partial x^i \partial x^j}(0).$$

Then, for all $\varepsilon > 0$, there is a φ in $\mathcal{D}(A) \cap C^2(\dot{\mathbb{R}}^\ell)$ with

$$\varphi(0) = \frac{\partial\varphi}{\partial x^i}(0) = 0, \qquad \frac{\partial^2\varphi}{\partial x^i \partial x^j}(0) = 2\delta_{ij}$$

and $\|\varphi-\psi\| \leq \varepsilon$. If ε is small enough, φ must be strictly positive on $\dot{\mathbb{R}}^\ell-\{0\}$. Fix such a φ, and let $\delta > 0$, $f \in C^2(\dot{\mathbb{R}}^\ell)$. By Theorem 5.1 again there is a g in $\mathcal{D}(A) \cap C^2(\dot{\mathbb{R}}^\ell)$ with

$$|f(y) - g(y)| \leq \delta\varphi(y)$$

for all y in $\dot{\mathbb{R}}^\ell$. Now

$$\frac{1}{t}\int |f(y) - g(y)| p^t(0,dy) \leq \frac{\delta}{t}\int \varphi(y) p^t(0,dy)$$

and since $\varphi \in \mathcal{D}(A)$ with $\varphi(0)=0$, the right hand side is (δ).

Therefore

(1) $$\frac{1}{t}\int [f(y) - f(0)] p^t(0,dy)$$

and

(2) $$\frac{1}{t}\int [g(y)-g(0)]p^t(0,dy)$$

differ by (δ). Since $g \in \mathcal{D}(A)$, (2) has a limit as $t \to 0$. Since δ is arbitrary, (1) has a limit as $t \to 0$. Therefore (1) is bounded as $t \to 0$. Since this is true for each f in the Banach space $C^2(\dot{\mathbb{R}}^\ell)$, by the principle of uniform boundedness there is a constant K such that for all f in $C^2(\dot{\mathbb{R}}^\ell)$ and $t > 0$,

$$\left|\frac{1}{t}(P^t f - f)(0)\right| \leq K\|f\|'.$$

By translation invariance,

$$\left\|\frac{1}{t}(P^t f - f)\right\| \leq K\|f\|'.$$

Now $\frac{1}{t}(P^t g - g) \to Ag$ for all g in the dense set $\mathcal{D}(A') \cap C^2(\dot{\mathbb{R}}^\ell)$, so by the Banach-Steinhaus theorem, $\frac{1}{t}(P^t f - f)$ converges in $C(\dot{\mathbb{R}}^\ell)$ for all f in $C^2(\dot{\mathbb{R}}^\ell)$. QED.

Theorem 5.3. Let P^t be a Markovian semigroup on $C(\dot{\mathbb{R}}^\ell)$, not necessarily commuting with translations, such that $C^2_{com}(\mathbb{R}^\ell) \subset \mathcal{D}(A)$, where A is the infinitesimal generator of P^t. If for all x in $\mathbb{R}^\ell$ and all $\varepsilon < 0$

(3) $$p^t(x,\{y : |y-x| \geq \varepsilon\}) = o(t),$$

then

(4) $$Af(x) = \sum_{i=1}^{\ell} b^i(x)\frac{\partial}{\partial x^i}f(x) + \sum_{i,j=1}^{\ell} a^{ij}(x)\frac{\partial^2}{\partial x^i \partial x^j}f(x)$$

for all f in $C^2_{com}(\mathbb{R}^\ell)$, where the a^{ij} and b^i are real and continuous, and for each x the matrix $a^{ij}(x)$ is of positive type.

A matrix a^{ij} is of positive type (positive definite, positive semi-definite, non-negative definite, etc.) in case for all complex ζ_i,

$$\sum_{i,j=1}^{\ell} \bar{\zeta}_i a^{ij} \zeta_j \geq 0.$$

The operator A is not necessarily elliptic, since the matrix $a^{ij}(x)$ may be singular. If P^t commutes with translations then a^{ij} and b^i are constants, of course.

<u>Proof.</u> Let $f \in C^2_{com}(\mathbb{R}^\ell)$ and suppose that f together with its first and second order partial derivatives vanishes at x. Let $g \in C^2_{com}(\mathbb{R}^\ell)$ be such that $g(y)=|y-x|^2$ in a neighborhood of x and $g \geq 0$. Let $\varepsilon > 0$ and let $U=\{y: |f(y)| \leq \varepsilon g(y)\}$ so that U is a neighborhood of x. By (3), $p^t(x,\mathbb{R}^\ell - U)=o(t)$ and so

$$Af(x) = \lim_{t \to 0} \frac{1}{t} \int f(y) p^t(x,dy)$$

$$= \lim_{t \to 0} \frac{1}{t} \int_U f(y) p^t(x,dy) \leq \varepsilon \lim_{t \to 0} \frac{1}{t} \int g(y) p^t(x,dy) = \varepsilon Ag(x).$$

Since ε is arbitrary, $Af(x)=0$. This implies that $Af(x)$ is of the form (4) for certain real numbers $a^{ij}(x)$, $b^i(x)$, and we may assume that the $a^{ij}(x)$ are symmetric. (There is no zero-order term since P^t is Markovian.) If we apply A to functions in $C^2_{com}(\mathbb{R}^\ell)$ which in a neighborhood of x agree with $y^i - x^i$ and $(y^i - x^i)(y^j - x^j)$, we see that b^i and a^{ij} are continuous. If f is in $C^2_{com}(\mathbb{R}^\ell)$ and $f(x)=0$ then

$$Af^2(x) = \lim_{t \to 0} \frac{1}{t} \int f^2(y) p^t(x,dy) \geq 0.$$

Therefore

$$Af^2(x) = \sum_{i,j=1}^{\ell} a^{ij}(x) \frac{\partial^2 f^2}{\partial x^i \partial x^j}(x)$$

$$= 2\sum_{i,j=1}^{\ell} a^{ij}(x) \frac{\partial f}{\partial x^i}(x) \frac{\partial f}{\partial x^j}(x) \geq 0.$$

We may choose

$$\frac{\partial f}{\partial x^i}(x) = \xi^i$$

to be arbitrary real numbers, and since $a^{ij}(x)$ is real and symmetric, $a^{ij}(x)$ is of positive type. QED.

<u>Theorem 5.4.</u> <u>Let</u> P^t <u>be a Markovian semigroup on</u> $C(\mathbb{R}^\ell)$ <u>commuting with translation, and let</u> A <u>be its infinitesimal generator.</u> <u>Then</u>

$$C^2(\mathbb{R}^\ell) \subset \mathcal{D}(A) \tag{5}$$

<u>and</u> P^t <u>is determined by</u> A <u>on</u> $C^2_{com}(\mathbb{R}^\ell)$.

<u>Proof.</u> The inclusion (5) follows from Theorem 5.2. The proof of that theorem shows that A is continuous from $C^2(\mathbb{R}^\ell)$ into $C(\mathbb{R}^\ell)$, so that A on $C^2_{com}(\mathbb{R}^\ell)$ determines A on $C^2(\mathbb{R}^\ell)$ by continuity. Since P^t commutes with translations, P^t leaves $C^2(\mathbb{R}^\ell)$ invariant.

Let $\lambda > 0$. We shall show that $(\lambda - A)C^2(\mathbb{R}^\ell)$ is dense in $C(\mathbb{R}^\ell)$. Suppose not. Then there is a non-zero continuous linear functional z on $C(\mathbb{R}^\ell)$ such that $(z,(\lambda-A)f)=0$ for all f in $C^2(\mathbb{R}^\ell)$. Since $C^2(\mathbb{R}^\ell)$ is dense in $C(\mathbb{R}^\ell)$, there is a g in $C^2(\mathbb{R}^\ell)$ with $(z,g)\neq 0$. Then

$$\frac{d}{dt}(z,P^tg) = (z,AP^tg) = (z,\lambda P^tg) = \lambda(z,P^tg)$$

since $P^t g$ is again in $C^2(\mathbb{R}^\ell)$. Therefore

$$(z, P^t g) = e^{\lambda t}(z, g)$$

is unbounded, which is a contradiction. It follows that if Q^t is another such semigroup with infinitesimal generator B, and B=A on $C^2_{com}(\mathbb{R}^\ell)$, then $(\lambda\text{-}B)^{-1}=(\lambda\text{-}A)^{-1}$ for $\lambda > 0$. But these are the Laplace transforms of the semigroups Q^t and P^t, and by the uniqueness theorem for Laplace transforms, $Q^t=P^t$. QED.

Theorem 5.5 follows from theorems 5.2, 5.3, 5.4 and the well-known formula for the fundamental solution of the diffusion equation.

References

[13]. G. A. Hunt, Semi-groups of measures on Lie groups, Transactions of the American Mathematical Society 81(1956), 264-293. (Hunt treats non-local processes as well, on arbitrary Lie groups.)

Banach spaces, the principle of uniform boundedness, the Banach-Steinhaus theorem, semigroups and infinitesimal generators are all discussed in detail in:

[14]. Einar Hille and Ralph S. Phillips, Functional Analysis and Semi-Groups, revised edition, American Math. Soc. Colloquium Publications, vol. XXXI, 1957.

§6. Gaussian processes

Gaussian random variables were discussed by Gauss in 1809 and the central limit theorem was stated by Laplace in 1812. Laplace had already considered Gaussian random variables around 1780, and for this reason Frenchmen call Gaussian random variables "Laplacean." However, the Gaussian measure and an important special case of the central limit theorem were discovered by De Moivre in 1733. The main tool in De Moivre's work was Stirling's formula, which, except for the fact that the constant occurring in it is $\sqrt{2\pi}$, was discovered by De Moivre. In statistical mechanics the Gaussian distribution is called "Maxwellian." Another name for it is "normal."

A <u>Gaussian measure</u> on $\mathbb{R}^{\ell}$ is a measure which is the transform of the measure with density

$$\frac{1}{(2\pi)^{\ell/2}}\, e^{-\frac{1}{2}|x|^2}$$

under an affine transformation. It is called <u>singular</u> in case the affine transformation is singular, which is the case if and only if it is singular with respect to Lebesgue measure.

A set of random variables is called Gaussian in case the distribution of each finite subset is Gaussian. A set of linear combinations, or limits in measure of linear combinations, of Gaussian random variables is Gaussian. Two (jointly) Gaussian random variables are independent if and only if they are uncorrelated; i.e., their covariance

$$r(x,y) = E(x-Ex)(y-Ey)$$

is zero (where E denotes the expectation).

§6. GAUSSIAN PROCESSES

We define the <u>mean</u> m and <u>covariance</u> r of a probability measure μ on $\mathbb{R}^\ell$ as follows, provided the integrals exist:

$$m_i = \int x_i \mu(dx)$$
$$r_{ij} = \int x_i x_j \mu(dx) - m_i m_j = \int (x_i - m_i)(x_j - m_j)\mu(dx)$$

where x has the components x_i. The covariance matrix r is of positive type. Let μ be a probability measure on $\mathbb{R}^\ell$, $\hat{\mu}$ its inverse Fourier transform

$$\hat{\mu}(\xi) = \int e^{i\xi\cdot x}\mu(dx).$$

Then μ is Gaussian if and only if

$$\hat{\mu}(\xi) = e^{-\frac{1}{2}\Sigma r_{ij}\xi_{ij} + i\Sigma m_i \xi_i}$$

in which case r is the covariance and m the mean. If r is nonsingular and r^{-1} denotes the inverse matrix, then the Gaussian measure with mean m and covariance r has the density

$$\frac{1}{(2\pi)^{\ell/2}(\det r)^{\frac{1}{2}}} e^{-\frac{1}{2}\Sigma(r^{-1})_{ij}(x_i - m_i)(x_j - m_j)}.$$

If r is of positive type there is a unique Caussian measure with covariance r and mean m.

A set of complex random variables is called Gaussian if and only if the real and imaginary parts are (jointly) Gaussian. We define the covariance of complex random variables by

$$r(x,y) = E(\overline{x} - E\overline{x})(y - Ey).$$

Let T be a set. A complex function r on T×T is called of <u>positive type</u> in case for all $t_1, \ldots, t_\ell$ in T the matrix $r(t_i, t_j)$ is of positive type. Let x be a stochastic process indexed by T. We call r(t,s)=r(x(t),x(s)) the <u>covariance</u> of the process, m(t)=Ex(t) the <u>mean</u> of the process (provided the integrals exist). The covariance is of positive type.

The following theorem is immediate (given the basic existence theorem for stochastic processes with prescribed finite joint distributions).

<u>Theorem 6.1.</u> <u>Let</u> T <u>be a set,</u> m <u>a function on</u> T, r <u>a function of positive type on</u> T×T. <u>Then there is a Gaussian stochastic process indexed by</u> T <u>with mean</u> m <u>and covariance</u> r. <u>Any two such are equivalent.</u>

Reference

[15]. J. L. Doob, Stochastic Processes, John Wiley & Sons, Inc., New York, 1953. (Gaussian processes are discussed on pp. 71-78.)

§7. The Wiener integral

The differences of the Wiener process

$$w(t)-w(s), \qquad 0 \leq s \leq t < \infty$$

form a Gaussian stochastic process, indexed by pairs of positive numbers s and t with $s \leq t$. This difference process has mean 0 and covariance

$$E(w(t)-w(s))(w(t')-w(s')) = \sigma^2 |[s,t] \cap [s',t']|$$

where $| \ |$ denotes Lebesgue measure, and σ^2 is the variance parameter of the Wiener process.

We may extend the difference process to all pairs of real numbers s and t. We may arbitrarily assign a distribution to w(0). The resulting stochastic process $w(t)$, $-\infty < t < \infty$, is called the two-sided Wiener process. It is Gaussian if and only if w(0) is Gaussian (e.g. $w(0)=x_0$ where x_0 is a fixed point), but in any case the differences are Gaussian. If we know that a Brownian particle is at x_0 at the present moment, $w(0)=x_0$, then w(t) for $t > 0$ is the position of the particle at time t in the future and w(t) for $t < 0$ is the position of the particle at time t in the past. A movie of Brownian motion looks, statistically, the same if it is run backwards.

We recall that, with probability one, the sample paths of the Wiener process are continuous but not differentiable. Nevertheless, integrals of the form

$$\int_{-\infty}^{\infty} f(t)dw(t)$$

can be defined, for any square-integrable f.

Theorem 7.1. Let Ω be the probability space of the differences of the two-sided Wiener process. There is a unique isometric operator from $L^2(\mathbb{R},\sigma^2 dt)$ into $L^2(\Omega)$, denoted

$$f \to \int_{-\infty}^{\infty} f(t)dw(t),$$

such that for all $-\infty < a \leq b < \infty$,

$$\int_{-\infty}^{\infty} \varphi_{[a,b]}(t)dw(t) = w(b)-w(a).$$

The set of $\int_{-\infty}^{\infty} f(t)dw(t)$ is Gaussian.

If E is any set, φ_E is its characteristic function,

$$\varphi_E(t) = \begin{cases} 1, & t \in E \\ 0, & t \notin E. \end{cases}$$

We shall write, in the future, $\int_a^b f(t)dw(t)$ for $\int_{-\infty}^{\infty} \varphi_{[a,b]}(t)f(t)dw(t)$.

Proof. Let f be a step function

$$f = \sum_{i=1}^{n} c_i \varphi_{[a_i,b_i]}.$$

Then we define

(1) $$\int_{-\infty}^{\infty} f(t)dw(t) = \sum_{i=1}^{n} c_i[w(b_i)-w(a_i)].$$

If g also is a step function,

$$g = \sum_{j=1}^{m} d_j \varphi_{[e_j,f_j]},$$

then

$$E \int_{-\infty}^{\infty} f(t)dw(t) \int_{-\infty}^{\infty} g(s)dw(s)$$

$$= E \sum_{i=1}^{n} c_i[w(b_i)-w(a_i)] \sum_{j=1}^{m} d_j[w(f_j)-w(e_j)]$$

$$= \sum_{i=1}^{n} \sum_{j=1}^{m} c_i d_j \sigma^2 |[a_i,b_i] \cap [e_j,f_j]|$$

$$= \sigma^2 \int_{-\infty}^{\infty} f(t)g(t)dt.$$

Since the step functions are dense in $L^2(\mathbb{R},\sigma^2 dt)$, the mapping extends by continuity to an isometry. Uniqueness is clear, and so is the fact that the random variables are Gaussian. QED.

The Wiener integral may be generalized. Let T,μ be an arbitrary measure space, and let $\mathcal{S}_0$ denote the family of measurable sets of finite measure. Let w be the Gaussian stochastic process indexed by $\mathcal{S}_0$ with mean 0 and covariance $r(E,F)=\mu(E\cap F)$. This is easily seen to be of positive type (see below). Let Ω be the probability space of the w-process.

Theorem 7.2. *There is a unique isometric mapping*

$$f \to \int f(t)dw(t)$$

from $L^2(T,\mu)$ *into* $L^2(\Omega)$ *such that, for* $E \in \mathcal{S}_0$,

$$\int \varphi_E(t)dw(t) = w(E).$$

The $\int f(t)dw(t)$ *are Gaussian.*

The proof is as before.

If $\mathcal{H}$ is a Hilbert space, the function r on $\mathcal{H}\times\mathcal{H}$ which is the inner product, $r(f,g)=(f,g)$, is of positive type, since

$$\Sigma\, \overline{\zeta}_i (f_i, f_j) \zeta_j = \|\sum_j \zeta_j f_j\|^2 \geq 0.$$

Consequently, the Wiener integral can be generalized further, as a purely Hilbert space-theoretic construct.

Theorem 7.3. *Let $\mathcal{H}$ be a Hilbert space. Then there is a Gaussian stochastic process, unique up to equivalence, with mean 0 and covariance given by the inner product.*

Proof. This follows from Theorem 6.1.

The special feature of the Wiener integral on the real line which makes it useful is its relation to differentiation.

Theorem 7.4. *Let f be of bounded variation on the real line with compact support, and let w be a Wiener process. Then*

$$\int_{-\infty}^{\infty} f(t)dw(t) = -\int_{-\infty}^{\infty} df(t)w(t). \tag{2}$$

In particular, if f is absolutely continuous on [a,b], then

$$\int_a^b f(t)dw(t) = -\int_a^b f'(t)w(t)dt + f(b)w(b) - f(a)w(a).$$

The left hand side of (2) is defined since f must be in L^2. The right hand side is defined a.e. (with probability one, that is) since almost every sample function of the Wiener process is continuous. The equality in (2) means equality a.e., of course.

Proof. If f is a step function, (2) is the definition (1) of the Wiener integral. In the general case we may let f_n be a sequence of step functions such that $f_n \to f$ in L^2 and $df_n \to df$ in the weak-* topology of measures, so that we have convergence to the two sides of (2). QED.

References

See Doob's book [15, §6, p.426] for a discussion of the Wiener integral. The purely Hilbert space approach to the Wiener integral, together with applications, has been developed by Irving Segal and others. See the following and its bibliography:

[16]. Irving Segal, Algebraic integration theory, Bulletin American Math. Soc. 71(1965), 419-489.

For discussions of Wiener's work see the special commemorative issue:

[17]. Bulletin American Math. Soc. 72(1966), No.1 Part 2.

We are assuming a knowledge of the Wiener process. For an exposition of the simplest facts see Appendix A of:

[18]. Edward Nelson, Feynman integrals and the Schrödinger equation, Journal of Mathematical Physics 5(1964), 332-343.

For an account of deeper facts, see:

[19]. Kiyosi Itô and Henry P. McKean, Jr., Diffusion Processes and their Sample Paths, Die Grundlehren der Mathematischen Wissenschaften in Einzeldarstellungen vol. 125, Academic Press, Inc., New York, 1965.

§8. A class of stochastic differential equations

By a Wiener process on $\mathbb{R}^\ell$ we mean a Markoff process w whose infinitesimal generator C is of the form

$$(1) \qquad C = \sum_{i,j=1}^{\ell} c^{ij} \frac{\partial^2}{\partial x^i \partial x^j},$$

where c^{ij} is a constant real matrix of positive type. Thus the $w(t)-w(s)$ are Gaussian, and independent for disjoint intervals, with mean 0 and covariance matrix $2c^{ij}|t-s|$.

THEOREM 8.1. *Let* $b: \mathbb{R}^\ell \to \mathbb{R}^\ell$ *satisfy a global Lipschitz condition; that is, for some constant* κ,

$$|b(x_0)-b(x_1)| \leq \kappa |x_0-x_1|,$$

for all x_0 *and* x_1 *in* $\mathbb{R}^\ell$. *Let* w *be a Wiener process on* $\mathbb{R}^\ell$ *with infinitesimal generator* C *given by (1). For each* x_0 *in* $\mathbb{R}^\ell$ *there is a unique stochastic process* $x(t)$, $0 \leq t < \infty$, *such that for all* t

$$(2) \qquad x(t) = x_0 + \int_0^t b(x(s))ds + w(t) - w(0).$$

The x *process has continuous sample paths with probability one.*

If we define $P^t f(x_0)$ *for* $0 \leq t < \infty$, $x_0 \in \mathbb{R}^\ell$, $f \in C(\mathbb{R}^\ell)$ *by*

$$(3) \qquad P^t f(x_0) = Ef(x(t)),$$

where E *denotes the expectation on the probability space of the* w *process, then* P^t *is a Markovian semigroup on* $C(\mathbb{R}^\ell)$. *Let* A *be the infinitesimal generator of* P^t. *Then* $C^2_{com}(\mathbb{R}^\ell) \subset \mathcal{D}(A)$ *and*

(4) $$Af = b\cdot\nabla f + Cf$$

for all f in $C^2_{com}(\mathbb{R}^\ell)$.

Proof. With probability one, the sample paths of the w process are continuous, so we need only prove existence and uniqueness for (2) with w a fixed continuous function of t. This is a classical result, even when w is not differentiable, and may be proved by the Picard method, as follows.

Let $\lambda > \kappa$, $t \geq 0$, and let $\mathfrak{X}$ be the Banach space of all continuous functions ξ from $[0,t]$ to $\mathbb{R}^\ell$ with the norm

$$\|\xi\| = \sup_{0 \leq s \leq t} e^{-\lambda s}|\xi(s)|.$$

Define the non-linear mapping $T: \mathfrak{X} \to \mathfrak{X}$ by

$$T\xi(s) = \xi(0) + \int_0^s b(x(r))dr + w(s) - w(0).$$

Then we have

(5) $$\begin{aligned}\|T\xi - T\eta\| &\leq |\xi(0)-\eta(0)| + \sup_{0\leq s\leq t} e^{-\lambda s}\left|\int_0^s [b(\xi(r)) - b(\eta(r))]dr\right| \\ &\leq |\xi(0)-\eta(0)| + \sup_{0\leq s\leq t} e^{-\lambda s}\kappa\int_0^s |\xi(r) - \eta(r)|dr \\ &\leq |\xi(0)-\eta(0)| + \sup_{0\leq s\leq t} e^{-\lambda s}\kappa\int_0^s e^{\lambda r}\|\xi-\eta\|dr \\ &= |\xi(0)-\eta(0)| + \alpha\|\xi-\eta\|,\end{aligned}$$

where $\alpha = \kappa/\lambda < 1$. For x_0 in $\mathbb{R}^\ell$, let $\mathfrak{X}_{x_0} = \{\xi \in \mathfrak{X} : \xi(0) = x_0\}$. Then $\mathfrak{X}_{x_0}$ is a complete metric space and by (5), T is a proper contraction on it. Therefore T has a unique fixed point x in $\mathfrak{X}_{x_0}$. Since t is arbitrary, there is a unique continuous function x from $[0,\infty)$ to $\mathbb{R}^\ell$

satisfying (2). Any solution of (2) is continuous, so there is a unique solution of (2).

Next we shall show that $P^t: C(\mathbb{R}^\ell) \to C(\mathbb{R}^\ell)$. By induction on (5),

$$\|T^n\xi - T^n\eta\| \leq [1+\alpha+\ldots+\alpha^{n-1}]\,|\xi(0)-\eta(0)| + \alpha^n\|\xi-\eta\|. \tag{6}$$

If x_0 is in $\mathbb{R}^\ell$, we shall also let x_0 denote the constant map $x_0(s)=x_0$, and we shall let x be the fixed point of T with $x(0)=x_0$, so that

$$x = \lim_{n\to\infty} T^n x_0,$$

and similarly for y_0 in $\mathbb{R}^\ell$. By (6), $\|x-y\| \leq \beta|x_0-y_0|$, where $\beta=1/(1-\alpha)$. Therefore, $|x(t)-y(t)| \leq e^{\lambda t}\beta|x_0-y_0|$. Now let f be any Lipschitz function on $\mathbb{R}^\ell$ with Lipschitz constant K. Then

$$|f(x(t)) - f(y(t))| \leq Ke^{\lambda t}\beta|x_0-y_0|.$$

Since this is true for each fixed w path, the estimate remains true when we take expectations, so that

$$|P^t f(x_0) - P^t f(y_0)| \leq Ke^{\lambda t}\beta|x_0-y_0|.$$

Therefore, if f is a Lipschitz function in $C(\mathbb{R}^\ell)$ then $P^t f$ is a bounded continuous function. The Lipschitz functions are dense in $C(\mathbb{R}^\ell)$ and P^t is a bounded linear operator. Consequently, if f is in $C(\mathbb{R}^\ell)$ then $P^t f$ is a bounded continuous function. We still need to show that it vanishes at infinity. By uniqueness,

$$x(t) = x(s) + \int_s^t b(x(r))dr + w(t) - w(s)$$

for all $0 \leq s \leq t$, so that

$$\begin{aligned}|x(t)-x(s)| &\le \left|\int_s^t [b(x(r))-b(x(t))]dr\right| + (t-s)|b(x(t))| + |w(t)-w(s)| \\ &\le \kappa \int_s^t |x(r)-x(t)|dr + t|b(x(t))| + |w(t)-w(s)| \\ &\le \kappa t \sup_{0\le r\le t} |x(r)-x(t)| + t|b(x(t))| + \sup_{0\le r\le t} |w(t)-w(r)|.\end{aligned}$$

Since this is true for each s, $0 \le s \le t$,

$$\sup_{0\le s\le t} |x(t)-x(s)| \le \gamma[t|b(x(t))| + \sup_{0\le s\le t} |w(t)-w(s)|],$$

where $\gamma = 1/(1-\kappa t)$, provided that $\kappa t < 1$. In particular, if $\kappa t < 1$ then

(7) $$|x(t)-x_0| \le \gamma[t|b(x(t))| + \sup_{0\le s\le t} |w(t)-w(s)|].$$

Now let f be in $C_{com}(\mathbb{R}^\ell)$, let $\kappa t < 1$, and let δ be the supremum of $|b(z_0)|$ for z_0 in the support of f. By (7), $f(x(t))=0$ unless

(8) $$\inf_{z_0 \in \text{supp } f} |z_0 - x_0| \le \gamma[t\delta + \sup_{0\le s\le t} |w(t)-w(s)|].$$

But as x_0 tends to infinity, the probability that w will satisfy (8) tends to 0. Since f is bounded, this means that $Ef(x(t)) = P^t f(x_0)$ tends to 0 as x_0 tends to infinity. We have already seen that $P^t f$ is continuous, so $P^t f$ is in $C(\mathbb{R}^\ell)$. Since $C_{com}(\mathbb{R}^\ell)$ is dense in $C(\mathbb{R}^\ell)$ and P^t is a bounded linear operator, P^t maps $C(\mathbb{R}^\ell)$ into itself, provided $\kappa t < 1$. This restriction could have been avoided by introducing an exponential factor, but this is not necessary, as we shall show that the P^t form a semigroup.

Let $0 \le s \le t$. The conditional distribution of $x(t)$, with $x(r)$ for all $0 \le r \le s$ given, is a function of $x(s)$ alone, since the equation

$$x(t) = x(s) + \int_s^t b(x(s'))ds' + w(t) - w(s)$$

has a unique solution. Thus the x process is a Markoff process, and

$$E\{f(x(t))|x(r),\ 0 \le r \le s\} = E\{f(x(t))|x(s)\} = P^{t-s}f(x(s))$$

for f in $C(\mathbb{R}^\ell)$, $0 \le s \le t$. Therefore,

$$\begin{aligned} P^{t+s}f(x_0) &= Ef(x(t+s)) \\ &= EE\{f(x(t+s))|x(r),\ 0 \le r \le s\} \\ &= EP^t f(x(s)) \\ &= P^s P^t f(x_0), \end{aligned}$$

so that $P^{t+s} = P^t P^s$. It is clear that

$$\sup_{0 \le f \le 1} P^t f(x_0) = 1$$

for all x_0 and t.

It remains only to prove (4) for f in $C^2_{com}(\mathbb{R}^\ell)$. (Since $C^2_{com}(\mathbb{R}^\ell)$ is dense in $C(\mathbb{R}^\ell)$ and the P^t have norm one, this will imply that $P^t f \to f$ as $t \to 0$ for all f in $C(\mathbb{R}^\ell)$, so that P^t is a Markovian semigroup.)

Let f be in $C^2_{com}(\mathbb{R}^\ell)$, and let K be a compact set containing the support of f in its interior. An argument entirely analogous to the derivation of (7), with the subtraction and addition of $b(x_0)$ instead of $b(x(t))$, gives

$$(9) \qquad |x(t) - x_0| \le \gamma[t|b(x_0)| + \sup_{0 \le s \le t} |w(0) - w(s)|],$$

provided $\kappa t < 1$ (which we shall assume to be the case). Let x_0 be in the complement of K. Then $f(x_0)=0$ and $f(x(t))$ is also 0 unless $\varepsilon \leq |x(t)-x_0|$, where ε is the distance from the support of f to the complement of K. But the probability that the right hand side of (9) will be bigger than ε is $o(t)$ (in fact, $o(t^n)$ for all n) by familiar properties of the Wiener process. Since f is bounded, this means that $P^t f(x_0)$ is uniformly $o(t)$ for x_0 in the complement of K, so that

$$\frac{P^t f(x_0) - f(x_0)}{t} \rightarrow b \cdot \nabla f(x_0) + Cf(x_0) = 0$$

uniformly for x_0 in the complement of K. Now let x_0 be in K. We have

$$P^t f(x_0) = Ef(x(t)) = Ef(x_0 + \int_0^t b(x(s))ds + w(t) - w(0)).$$

Define R(t) by

$$\begin{aligned} f(x_0 &+ \int_0^t b(x(s))ds + w(t) - w(0)) \\ &= f(x_0) + tb(x_0) \cdot \nabla f(x_0) + [w(t) - w(0)] \cdot \nabla f(x_0) \\ &\quad + \frac{1}{2} \sum_{i,j} [w^i(t) - w^i(0)][w^j(t) - w^j(0)] \frac{\partial^2}{\partial x^i \partial x^j} f(x_0) + R(t). \end{aligned}$$

Then

$$\frac{P^t f(x_0) - f(x_0)}{t} = b(x_0) \cdot \nabla f(x_0) + Cf(x_0) + \frac{1}{t} ER(t).$$

By Taylor's formula,

$$R(t) = o(|w(t) - w(0)|^2) + o(\int_0^t [b(x(s)) - b(x_0)]ds).$$

Since $E(|w(t)-w(0)|^2) \leq$ const. t, we need only show that

(10) $$E \sup_{x_0 \in K} \frac{1}{t} \int_0^t |b(x(s)) - b(x_0)| ds$$

tends to 0. But (10) is less than

$$E \sup_{x_0 \in K} \frac{1}{t} \kappa \int_0^t |x(s) - x_0| ds,$$

which by (9) is less than

(11) $$E \sup_{x_0 \in K} \kappa\gamma[t|b(x_0)| + \sup_{0 \leq s \leq t} |w(0) - w(s)|].$$

The integrand in (11) is integrable and decreases to 0 as $t \to 0$. QED.

Theorem 8.1 may be generalized in various ways. The first paragraph of the theorem remains true if b is a continuous function of x and t which satisfies a global Lipschitz condition in x with a uniform Lipschitz constant for each compact t-interval. The second paragraph needs to be slightly modified as we no longer have a semi-group, but the proofs are the same. Doob [15, §6, pp.273-291], using K. Itô's stochastic integrals (see §11), has a much deeper generalization in which the matrix c^{ij} depends on x and t. The restriction that b satisfy a global Lipschitz condition is necessary in general. For example, if the matrix c^{ij} is 0 then we have a system of ordinary differential equations. However, if C is elliptic (that is, if the matrix c^{ij} is of positive type and non-singular) the smoothness conditions on b may be greatly relaxed (cf.[20]).

We make the convention that

$$dx(t) = b(x(t))dt + dw(t)$$

mens that

$$x(t) - x(s) = \int_s^t b(x(r))dr + w(t) - w(s)$$

for all t and s.

<u>Theorem 8.2.</u> <u>Let</u> $A: \mathbb{R}^\ell \to \mathbb{R}^\ell$ <u>be linear, let</u> w <u>be a Wiener process on</u> $\mathbb{R}^\ell$ <u>with infinitesimal generator (1), and let</u> $f:[0,\infty) \to \mathbb{R}^\ell$ <u>be continuous. Then the solution of</u>

$$dx(t) = Ax(t)dt + f(t)dt + dw(t), \quad x(0) = x_0, \tag{12}$$

<u>for</u> $t \geq 0$ <u>is</u>

$$x(t) = e^{At}x_0 + \int_0^t e^{A(t-s)}f(s)ds + \int_0^t e^{A(t-s)}dw(s). \tag{13}$$

<u>The</u> x(t) <u>are Gaussian with mean</u>

$$Ex(t) = e^{At}x_0 + \int_0^t e^{A(t-s)}f(s)ds \tag{14}$$

<u>and covariance</u> $r(t,s)=E(x(t)-Ex(t))(x(s)-Ex(s))$ <u>given by</u>

$$r(t,s) = \begin{cases} e^{A(t-s)} \int_0^s e^{Ar}2ce^{A^T r}dr, & t \geq s \\ \int_0^t e^{Ar}2ce^{A^T r}dre^{A(s-t)}, & t \leq s. \end{cases} \tag{15}$$

The latter integral in (13) is a Wiener integral (as in §7). In (15), A^T denotes the transpose of A and c is the matrix with entries c^{ij} occurring in (1).

<u>Proof.</u> Define x(t) by (13). Integrate the last term in (13) by parts, obtaining

$$\int_0^t e^{A(t-s)}dw(s) = \int_0^t Ae^{A(t-s)}w(s)ds + e^{A(t-s)}w(s)\Big|_{s=0}^{s=t}$$

$$= \int_0^t Ae^{A(t-s)}w(s)ds + w(t) - e^{At}w(0).$$

It follows that $x(t)-w(t)$ is differentiable, and has derivative $Ax(t)+f(t)$. This proves that (12) holds.

The $x(t)$ are clearly Gaussian with the mean (14). Suppose that $t \geq s$. Then the covariance is given by

$$Ex_i(t)x_j(s) - Ex_i(t)Ex_j(s)$$

$$= E \int_0^t \sum_k (e^{A(t-t_1)})_{ik} dw_k(t_1) \int_0^s \sum_h (e^{A(s-s_1)})_{jh} dw_h(s_1)$$

$$= \int_0^s \sum_{k,h} (e^{A(t-r)})_{ik} 2c_{kh} (e^{A(s-r)})_{jh} dr$$

$$= \int_0^s (e^{A(t-r)} 2c e^{A^T(s-r)})_{ij} dr$$

$$= (e^{A(t-s)} \int_0^s e^{Ar} 2c e^{A^T r} dr)_{ij}.$$

The case $t \leq s$ is analogous. QED.

Reference

[20]. Edward Nelson, Les écoulements incompressibles d'énergie finie, Colloques internationaux du Centre national de la recherche scientifique N° 117, Les équations aux dérivées partielles, Editions du C.N.R.S., Paris, 1962. (The last statement in section II is incorrect.)

§9. The Ornstein-Uhlenbeck theory of Brownian motion

The theory of Brownian motion developed by Einstein and Smoluchowski, although in agreement with experiment, was clearly a highly idealized treatment. The theory was far removed from the Newtonian mechanics of particles. Langevin initiated a train of thought which, in 1930, culminated in a new theory of Brownian motion by L. S. Ornstein and G. E. Uhlenbeck [22]. For ordinary Brownian motion (e.g. carmine particles in water) the predictions of the Ornstein-Uhlenbeck theory are numerically indistinguishable from those of the Einstein-Smoluchowski theory. However, the Ornstein-Uhlenbeck theory is a truly dynamical theory and represents great progress in the understanding of Brownian motion. Also, as we shall see later (§10), there is a Brownian motion where the Einstein-Smoluchowski theory breaks down completely and the Ornstein-Uhlenbeck theory is successful.

The program of reducing Brownian motion to Newtonian particle mechanics is still incomplete. The problem, or one formulation of it, is to deduce each of the following theories from the one below it:

Einstein - Smoluchowski
Ornstein - Uhlenbeck
Maxwell - Boltzmann
Hamilton - Jacobi

We shall consider the first of these reductions in detail later (§10). Now we shall describe the Ornstein-Uhlenbeck theory for a free particle and compare it with Einstein's theory.

We let $x(t)$ denote the position of a Brownian particle at time t and assume that the velocity $dx/dt=v$ exists and satisfies the Langevin equation

(1) $$dv(t) = -\beta v(t)dt + dB(t).$$

Here B is a Wiener process (with variance parameter to be determined later) and β is a constant with the dimensions of frequency (inverse time). Let m be the mass of the particle, so that we may write

$$m\frac{d^2x}{dt^2} = -m\beta v + m\frac{dB}{dt}.$$

This is merely formal since B is not differentiable. Thus (using Newton's law $F=ma$) we are considering the force on a free Brownian particle as made up of two parts, a frictional force $F_0=-m\beta v$ with friction coefficient $m\beta$ and a fluctuating force $F_1=mdB/dt$ which is (formally) a Gaussian stationary process with correlation function of the form a constant times δ, where the constant will be determined later.

If $v(0)=v_0$ and $x(0)=x_0$ the solution of the initial value problem is, by Theorem 8.2,

(2) $$\begin{aligned} v(t) &= e^{-\beta t}v_0 + e^{-\beta t}\int_0^t e^{\beta s}dB(s), \\ x(t) &= x_0 + \int_0^t v(s)ds. \end{aligned}$$

For a free particle there is no loss of generality in considering only the case of one-dimensional motion. Let σ^2 be the variance parameter of B (infinitesimal generator $\frac{1}{2}\sigma^2 d^2/dv^2$, $EdB(t)^2=\sigma^2 dt$).

The velocity $v(t)$ is Gaussian with mean

$$e^{-\beta t}v_0,$$

by (2). To compute the covariance, let $t \geq s$. Then

$$E\left(e^{-\beta t}\int_0^t e^{\beta t_1} dB(t_1) e^{-\beta s}\int_0^s e^{\beta s_1} dB(s_1)\right)$$

$$= e^{-\beta(t+s)}\int_0^s e^{2\beta r}\sigma^2 dr$$

$$= e^{-\beta(t+s)}\sigma^2 \frac{e^{2\beta s}-1}{2\beta} .$$

For $t=s$ this is

$$\frac{\sigma^2}{2\beta}\left(1-e^{-2\beta t}\right).$$

Thus, no matter what v_0 is, the limiting distribution of $v(t)$ as $t \to \infty$ is Gaussian with mean 0 and variance $\sigma^2/2\beta$. Now the law of equipartition of energy in statistical mechanics says that the mean energy of the particle (in equilibrium) per degree of freedom should be $\frac{1}{2}kT$. Therefore we set

$$\frac{1}{2}m\,\frac{\sigma^2}{2\beta} = \frac{1}{2}kT.$$

That is, recalling the previous notation $D=kT/m\beta$, we adopt the notation

$$\sigma^2 = 2\,\frac{\beta kT}{m} = 2\beta^2 D$$

for the variance parameter of B.

We summarize in the following theorem.

<u>Theorem 9.1.</u> <u>Let</u> D <u>and</u> β <u>be strictly positive constants and let</u> B <u>be the Wiener process on</u> $\mathbb{R}$ <u>with variance parameter</u> $2\beta^2 D$.

The solution of

$$dv(t) = -\beta v(t)dt + dB(t); \qquad v(0) = v_0$$

for $t \geq 0$ *is*

$$v(t) = e^{-\beta t}v_0 + \int_0^t e^{-\beta(t-s)}dB(s).$$

The random variables $v(t)$ *are Gaussian with mean*

$$m(t) = e^{-\beta t}v_0$$

and covariance

$$r(t,s) = \frac{D}{\beta}\left(e^{-\beta|t-s|} - e^{-\beta(t+s)}\right).$$

The $v(t)$ *are the random variables of the Markoff process on* $\mathbb{R}$ *with infinitesimal generator*

$$-\beta v \frac{d}{dv} + \beta^2 D \frac{d^2}{dv^2}$$

with domain including $C^2_{com}(\mathbb{R})$, *with initial measure* δ_{v_0}. *The kernel of the corresponding semigroup operator* P^t *is given by*

$$p^t(v_0,dv) = [2\pi\beta D(1-e^{-2\beta t})]^{-\frac{1}{2}} \exp\left[-\frac{(v-e^{-\beta t}v_0)^2}{2\beta D(1-e^{-2\beta t})}\right]dv.$$

The Gaussian measure μ *with mean* 0 *and variance* βD *is invariant,* $P^{t*}\mu=\mu$, *and* μ *is the limiting distribution of* $v(t)$ *as* $t \to \infty$.

The process v is called the Ornstein-Uhlenbeck velocity process with diffusion coefficient D and relaxation time β^{-1}, and the corresponding position process x (given by (2)) is called the Ornstein-Uhlenbeck process.

Theorem 9.2. *Let the* $v(t)$ *be as in Theorem 9.1, and let*

$$x(t) = x_0 + \int_0^t v(s)ds.$$

Then the $x(t)$ *are Gaussian with mean*

$$\tilde{m}(t) = x_0 + \frac{1-e^{-\beta t}}{\beta} v_0$$

and covariance

$$\tilde{r}(t,s) = 2D \min(t,s) + \frac{D}{\beta}(-2 + 2e^{-\beta t} + 2e^{-\beta s} - e^{-\beta|t-s|} - e^{-\beta(t+s)}).$$

Proof. This follows from Theorem 9.1 by integration,

$$\tilde{m}(t) = x_0 + \int_0^t m(s)ds,$$

$$\tilde{r}(t,s) = \int_0^t dt_1 \int_0^s ds_1 r(t_1,s_1).$$

The second integration is tedious but straightforward. QED.

In particular, the variance of $x(t)$ is

$$2Dt + \frac{D}{\beta}(-3 + 4e^{-\beta t} - e^{-2\beta t}).$$

The variance in Einstein's theory is $2Dt$. By elementary calculus, the absolute value of the difference of the two variances is less than $3D\beta^{-1}$. In the typical case of $\beta^{-1}=10^{-8}$ sec., $t=\frac{1}{2}$ sec., we make a proportional error of less than 3×10^{-8} by adopting Einstein's value for the variance. The following theorem shows that the Einstein theory is a good approximation to the Ornstein-Uhlenbeck theory for a free particle.

Theorem 9.3. *Let* $0=t_0 < t_1 <\ldots< t_n$, *and let*

$$\Delta t = \min_{1\leq i\leq n} t_i - t_{i-1}.$$

Let $f(x_1,\ldots,x_n)$ *be the probability density function for* $x(t_1),\ldots,x(t_n)$, *where* x *is the Ornstein-Uhlenbeck process with* $x(0)=x_0$, $v(0)=v_0$, *diffusion coefficient* D *and relaxation time* β^{-1}. *Let* $g(x_1,\ldots,x_n)$ *be the probability density function for* $w(t_1),\ldots,w(t_n)$, *where* w *is the Wiener process with* $w(0)=x_0$ *and diffusion coefficient* D.

Let $\varepsilon > 0$. *There exist* N_1 *depending only on* ε *and* n *and* N_2 *depending only on* ε *such that if*

$$\Delta t \geq N_1 \beta^{-1}, \tag{3}$$

$$t_1 \geq N_2 \frac{v_0^2}{2D\beta^2}, \tag{4}$$

then

$$\int_{\mathbb{R}^n} |f(x_1,\ldots,x_n) - g(x_1,\ldots,x_n)|\, dx_1\ldots dx_n \leq \varepsilon. \tag{5}$$

Proof. Assume, as one may without loss of generality, that $x_0=0$. Consider the non-singular linear transformation $(x_1,\ldots,x_n) \to (\tilde{x}_1,\ldots,\tilde{x}_n)$ on $\mathbb{R}^n$ given by

$$\tilde{x}_i = [2D(t_i - t_{i-1})]^{-\frac{1}{2}}(x_i - x_{i-1}) \tag{6}$$

for $i=1,\ldots,n$. The random variables $\tilde{w}(t_i)$ obtained when this transformation is applied to the $w(t_i)$ are orthonormal since $Ew(t_i)w(t_j)=2D\min(t_i,t_j)$. Thus $\tilde{g}$, the probability density function of the $\tilde{w}(t_i)$, is the unit Gaussian function on $\mathbb{R}^n$. Let $\tilde{f}$ be the probability density function of the $\tilde{x}(t_i)$, where the $\tilde{x}(t_i)$ are obtained by applying the linear transformation (6) to the $x(t_i)$. The left hand side of (5) is unchanged if we replace f by $\tilde{f}$ and g by $\tilde{g}$, since the total variation norm of a measure is unchanged under a one-to-one measurability-preserving map such as (6).

We use the notation Cov for the covariance of two random variables, Cov $xy = Exy - ExEy$. By Theorem 9.2 and the remark following it,

$$\text{Cov } x(t_i)x(t_j) = \text{Cov } w(t_i)w(t_j) + \varepsilon_{ij},$$

where $|\varepsilon_{ij}| \leq 3D\beta^{-1}$. By (6),

$$\text{Cov } \tilde{x}(t_i)\tilde{x}(t_j) = \delta_{ij} + \varepsilon'_{ij},$$

where $|\varepsilon'_{ij}| \leq 4 \cdot 3D\beta^{-1}/2D\Delta t \leq 6/N_1$ if (3) holds. Again by Theorem 9.2, the mean of $\tilde{x}(t_1)$ is, in absolute value, smaller than

$$|v_0|/\beta[2Dt_1]^{\frac{1}{2}} \leq N_2^{\frac{1}{2}}$$

if (4) holds. The mean of $\tilde{x}(t_i)$ for $i>1$ is, in absolute value, smaller than

$$(e^{-\beta t_{i-1}} - e^{-\beta t_i})|v_0|/\beta[2D(t_i - t_{i-1})]^{\frac{1}{2}}.$$

Since the first factor is smaller than 1, the square of this is smaller than

$$\frac{e^{-\beta t_{i-1}} - e^{-\beta t_i}}{t_i - t_{i-1}} \frac{v_0^2}{2D\beta^2} \leq \frac{\beta t_1}{N_2} e^{-\beta t_1} \leq \frac{N_1 e^{-N_1}}{N_2}$$

if (3) and (4) hold with $N_1 \geq 1$. Therefore, if we choose N_1 and N_2 large enough, the mean and convariance of $\tilde{f}$ are arbitrarily close to 0 and δ_{ij}, respectively, which concludes the proof. QED.

Chandrasekhar omits the condition (4) in his discussion [21, equations (171) through (174)], but his reasoning is circular. Clearly, if v_0 is enormous then t_1 must be suitably large before the Wiener process is a good approximation. The condition (3) is usually written $\Delta t \gg \beta^{-1}$ (Δt much larger than β^{-1}). If v_0 is a typical velocity, i.e. if $|v_0|$ is not much larger than the standard deviation $(kT/m)^{\frac{1}{2}}=(D\beta)^{\frac{1}{2}}$ of the Maxwellian velocity distribution, then the condition (4), $t_1 \gg v_0^2/2D\beta^2$, is no additional restriction if $\Delta t \gg \beta^{-1}$.

There is another, and quite weak, formulation of the fact that the Wiener process is a good approximation to the Ornstein-Uhlenbeck process for a free particle in the limit of very large β (very short relaxation time) but D of reasonable size.

<u>Definition.</u> Let x_α, x be real stochastic processes indexed by the same index set T but not necessarily defined on a common probability space. We say that x_α <u>converges to</u> x <u>in distribution</u> in case for each $t_1,\ldots,t_n$ in T, the distribution of $x_\alpha(t_1),\ldots,x_\alpha(t_n)$ converges (in the weak-* topology of measures on $\mathbb{R}^n$, as α ranges over a directed set) to the distribution of $x(t_1),\ldots,x(t_n)$.

It is easy to see that if we represent all of the processes in the usual way [25] on $\Omega=\Pi\dot{\mathbb{R}}$, this is the same as saying that $\Pr_\alpha$ converges to Pr in the weak-* topology of regular Borel measures on Ω, where $\Pr_\alpha$ is the regular Borel measure associated with x_α and Pr is the regular measure associated with x.

The following two theorems are trivial.

Theorem 9.4. *Let x_α, x be Gaussian stochastic processes with means m_α, m and covariances r_α, r. Then x_α converges to x in distribution if and only if $r_\alpha \to r$ and $m_\alpha \to m$ pointwise (on T and $T \times T$ respectively, where T is the common index set of the processes).*

Theorem 9.5. *Let β and σ^2 vary in such a way that $\beta \to \infty$ and $D = \sigma^2/2\beta^2$ remains constant. Then for all v_0 the Ornstein-Uhlenbeck process with initial conditions $x(0)=x_0$, $v(0)=v_0$, diffusion coefficient D, and relaxation time β^{-1} converges in distribution to the Wiener process starting at x_0 with diffusion coefficient D.*

References

The best account of the Ornstein-Uhlenbeck theory and related matters is

[21]. S. Chandrasekhar, *Stochastic problems in physics and astronomy*, Reviews of Modern Physics 15(1943), 1-89.

See also

[22]. G.E. Uhlenbeck and L. S. Ornstein, *On the theory of Brownian motion*, Physical Review 36(1930), 823-841.

[23]. Ming Chen Wang and G. E. Uhlenbeck, *On the theory of Brownian motion II*, Reviews of Modern Physics 17(1945), 323-342.

The first mathematically rigorous treatment, and in addition the source of great conceptual and computational simplifications, was

[24]. J. L. Doob, The Brownian movement and stochastic equations, Annals of Mathematics 43(1942), 351-369.

All four of the above articles are reprinted in the Dover paperback "Selected Papers on Noise and Stochastic Processes," edited by Nelson Wax.

[25]. E. Nelson, Regular probability measures on function space, Annals of Mathematics 69(1959), 630-643.

§10. Brownian motion in a force field

We continue the discussion of the Ornstein-Uhlenbeck theory. Suppose we have a Brownian particle in an external field of force given by $K(x,t)$ in units of force per unit mass (acceleration). Then the Langevin equations of the Ornstein-Uhlenbeck theory become

$$\begin{aligned} dx(t) &= v(t)dt \\ dv(t) &= K(x(t),t)dt - \beta v(t)dt + dB(t), \end{aligned} \tag{1}$$

where B is a Wiener process with variance parameter $2\beta^2 D$. This is of the form considered in Theorem 8.1:

$$d\begin{pmatrix} x(t) \\ v(t) \end{pmatrix} = \begin{pmatrix} v(t) \\ K(x(t),t) - \beta v(t) \end{pmatrix} dt + d\begin{pmatrix} 0 \\ B(t) \end{pmatrix}.$$

Notice that we can no longer consider the velocity process, or a component of it, by itself.

For a free particle ($K=0$) we have seen that the Wiener process, which is a Markoff process on coordinate space (x-space) is a good approximation, except for very small time intervals, to the Ornstein-Uhlenbeck process, which is a Markoff process on phase space (x,v-space). Similarly, when an external force is present, there is a Markoff process on coordinate space, discovered by Smoluchowski, which under certain circumstances is a good approximation to the position $x(t)$ of the above Ornstein-Uhlenbeck process.

Suppose, to begin with, that K is a constant. The force on a particle of mass m is Km and the friction coefficient is $m\beta$, so the particle should acquire the limiting velocity $Km/m\beta = K/\beta$. That is, for times large compared to the relaxation time β^{-1} the velocity should be approximately K/β. If we include the random fluctuations due to Brownian motion, this suggests the equation

$$dx(t) = \frac{K}{\beta}\,dt + dw(t)$$

where w is the Wiener process with diffusion coefficient $D = kT/m\beta$. (If there were no diffusion we would have, approximately for $t \gg \beta^{-1}$, $dx(t) = (K/\beta)dt$, and if there were no force we would have $dx(t) = dw(t)$.) If now K depends on x and t, but varies so slowly that it is approximately constant along trajectories for times of the order β^{-1}, we write

$$dx(t) = \frac{K(x(t),t)}{\beta}\,dt + dw(t).$$

This is the basic equation of the Smoluchowski theory; cf. Chandrasekhar's discussion [21].

We shall begin by discussing the simplest case, when K is linear and independent of t. Consider the one-dimensional harmonic oscillator with circular frequency ω. The Langevin equation in the Ornstein-Uhlenbeck theory is then

$$\begin{aligned} dx(t) &= v(t)dt \\ dv(t) &= -\omega^2 x(t)dt - \beta v(t)dt + dB(t) \end{aligned}$$

or

$$d\begin{pmatrix} x \\ v \end{pmatrix} = \begin{pmatrix} 0 & 1 \\ -\omega^2 & -\beta \end{pmatrix}\begin{pmatrix} x \\ v \end{pmatrix}dt + d\begin{pmatrix} 0 \\ B \end{pmatrix},$$

where, as before, B is a Wiener process with variance parameter

$\sigma^2 = 2\beta kT/m = 2\beta^2 D$.

The characteristic equation of the matrix

$$A = \begin{pmatrix} 0 & 1 \\ -\omega^2 & -\beta \end{pmatrix}$$

is $\mu^2 + \beta\mu + \omega^2 = 0$, with the eigenvalues

$$\mu_1 = -\tfrac{1}{2}\beta + \sqrt{\tfrac{1}{4}\beta^2 - \omega^2}, \qquad \mu_2 = -\tfrac{1}{2}\beta - \sqrt{\tfrac{1}{4}\beta^2 - \omega^2}\ .$$

As in the elementary theory of the harmonic oscillator without Brownian motion, we distinguish three cases:

overdamped	$\beta > 2\omega$,
critically damped	$\beta = 2\omega$,
underdamped	$B < 2\omega$.

Except in the critically damped case, the matrix $\exp(tA)$ is

$$e^{tA} = \frac{1}{\mu_2 - \mu_1}\begin{pmatrix} \mu_2 e^{\mu_1 t} - \mu_1 e^{\mu_2 t} & -e^{\mu_1 t} + e^{\mu_2 t} \\ \mu_2\mu_1 e^{\mu_1 t} - \mu_2\mu_1 e^{\mu_2 t} & -\mu_1 e^{\mu_1 t} + \mu_2 e^{\mu_2 t} \end{pmatrix}.$$

(We derive this as follows. Each matrix entry must be a linear combination of $\exp(\mu_1 t)$ and $\exp(\mu_2 t)$. The coefficients are determined by the requirements that $\exp(tA)$ and $d\,\exp(tA)/dt$ are 1 and A respectively when $t=0$.)

We let $x(0)=x_0$, $v(0)=v_0$. Then the mean of the process is

$$e^{tA}\begin{pmatrix} x_0 \\ v_0 \end{pmatrix}.$$

The covariance matrix of the Wiener process $\binom{0}{B}$ is

$$2c = \begin{pmatrix} 0 & 0 \\ 0 & 2\beta^2 D \end{pmatrix}.$$

The covariance matrix of the x,v process can be determined from Theorem 8.2, but the formulas are complicated and not very illuminating. The covariances for equal times are listed by Chandrasekhar [21, original page 30].

The Smoluchowski approximation is

$$dx(t) = -\frac{\omega^2}{\beta} x(t)dt + dw(t),$$

where w is a Wiener process with diffusion coefficient D. This has the same form as the Ornstein-Uhlenbeck velocity process for a free particle. According to the intuitive argument leading to the Smoluchowski equation, it should be a good approximation for time intervals large compared to the relaxation time ($\Delta t \gg \beta^{-1}$) when the force is slowly varying ($\beta \gg 2\omega$; i.e., the highly overdamped case).

The Brownian motion of a harmonically bound particle has been investigated experimentally by Gerlach and Lehrer and by Kappler [26]. The particle is a very small mirror suspended in a gas by a thin quartz fiber. The mirror can rotate but the torsion of the fiber supplies a linear restoring force. Bombardment of the mirror by the molecules of the gas causes a Brownian motion of the mirror. The Brownian motion is one dimensional, being described by the angle which the mirror makes with its equilibrium position. (This angle, which is very small, can be measured accurately by shining a light on the mirror and measuring the position of the reflected spot a large distance away.) At atmospheric pressure the motion is highly overdamped, but at sufficiently low pressures the underdamped case can be observed, too. The Ornstein-Uhlenbeck theory gives for the invariant measure (limiting distribution as $t \to \infty$) $Ex^2=kT/m\omega^2$ and $Ev^2=kT/m$. That is, the expected value of the kinetic energy $\frac{1}{2}mv^2$ in equilibrium is $\frac{1}{2}kT$, in accordance with the equipartition law of statistical mechanics. These values are independent of β, and the constancy of Ex^2 as the pressure varies was observed experimentally. However, the appearance of the trajectories varies tremendously.

Consider Fig. 5a on p.243 of Kappler [26], which is the same as Fig. 5b on p.169 of Barnes and Silverman [11, §3]. This is a record of the motion in the highly overdamped case. Locally the graph looks very much like the Wiener process, extremely rough. However, the graph never rises very far above or sinks very far below a median position, and there is a general tendency to return to the median position. If we reverse the direction of time, the graph looks very much the same. This process is a Markoff process - there is no memory of previous positions. A graph of the velocity in the Ornstein-Uhlenbeck process for a free particle would look the same.

Now consider Fig. 6a on p.244 of Kappler (Fig. 5c on p.169 of Barnes and Silverman). This is a record of the motion in the underdamped case. The curve looks smooth and more or less sinusoidal. This is clearly not the graph of a Markoff process, as there is an evident distinction between the upswings and downswings of the curve. Consequently, the Smoluchowski approximation is completely invalid in this case. When Barnes and Silverman reproduced the graph, it got turned upside down, reversing the direction of time. However, the over-all appearance of the curves is very much the same and in fact this stochastic process is invariant under time reversal. Are there any beats in this graph, and should there be?

Fig. 4b on p.242 of Kappler (fig. 5a on p.169 of Barnes and Silverman) represents an intermediate case.

We illustrate crudely the two cases (Fig. 2). Fig. 2a is the highly overdamped case, a Markoff process. Fig. 2b is the underdamped case, not a Markoff process. Fig. 2c illustrates a case which does not occur. (The only repository for memory is in the velocity, so over-all sinusoidal behavior implies local smoothness of the curve.)

One has the feeling with some of Kappler's curves that one can occasionally see where an exceptionally energetic gas molecule gave the mirror a kick. This is not true. Even at the lowest pressure used, an enormous number of collisions takes place per period, and the irregularities in the curves are due to chance fluctuations in the sum of enormous numbers of individually negligible events.

It is not correct to think simply that the jiggles in a Brownian trajectory are due to kicks from molecules. Brownian motion is unbelievably gentle. Each collision has an entirely negligible effect on the position of the Brownian particle, and

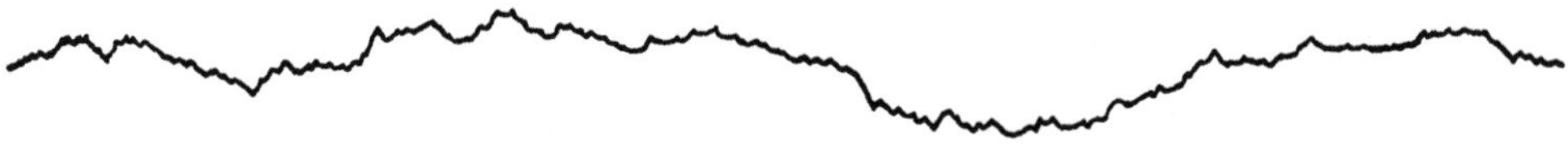

Fig. 2a

Fig. 2b

Fig. 2c

it is only fluctuations in the accumulation of an enormous number of very slight changes in the particle's velocity which give the trajectory its irregular appearance.

The experimental results lend credence to the statement that the Smoluchowski approximation is valid when the friction is large (β large). A theoretical proof does not seem to be in the literature. Ornstein and Uhlenbeck [22] show only that if a harmonically bound particle starts at x_0 at time 0 with a Maxwellian-distributed velocity, the mean and variance of $x(t)$ are approximately the mean and variance of the Smoluchowski theory provided $\beta \gg 2\omega$ and $t \gg \beta^{-1}$. We shall examine the Smoluchowski approximation in the case of a general external force, and prove a result which says that it is in a very strong sense the limiting case of the Ornstein-Uhlenbeck theory for large friction.

Consider the equations (1) of the Ornstein-Uhlenbeck theory. Let w be a Wiener process with diffusion coefficient D (variance parameter $2D$) as in the Einstein-Smoluchowski theory. Then if we set $B = \beta w$ the process B has the correct variance parameter $2\beta^2 D$ for the Ornstein-Uhlenbeck theory. The idea of the Smoluchowski approximation is that the relaxation time β^{-1} is negligibly small but that the diffusion coefficient $D = kT/m\beta$ and the velocity K/β are of significant size. Let us therefore define $b(x,t)$ (having the dimensions of a velocity) by

$$b(x,t) = \frac{K(x,t)}{\beta},$$

and study the solution of (1) as $\beta \to \infty$ with b and D fixed. The equations (1) become

$$dx(t) = v(t)dt$$
$$dv(t) = -\beta v(t)dt + \beta b(x(t),t)dt + \beta dw(t).$$

Let $x(t)=x(\beta,t)$ be the solution of these equations with $x(0)=x_0$, $v(0)=v_0$. We will show that as $\beta \to \infty$, $x(t)$ converges to the solution $y(t)$ of the Smoluchowski equation

$$dy(t) = -b(y(t),t)dt + dw(t)$$

with $y(0)=x_0$. For simplicity, we treat the case that b is independent of the time, although the theorem and its proof remain valid for the case that b is continuous and, for t in compact sets, satisfies a uniform Lipschitz condition in x.

Theorem 10.1. *Let* $b:\mathbb{R}^\ell \to \mathbb{R}^\ell$ *satisfy a global Lipschitz condition and let* w *be a Wiener process on* $\mathbb{R}^\ell$. *Let* x,v *be the solution of the coupled equations*

$$(2) \qquad dx(t) = v(t)dt \qquad ; \qquad x(0) = x_0,$$

$$(3) \qquad dv(t) = -\beta v(t)dt + \beta b(x(t))dt + \beta dw(t); \qquad v(0) = v_0.$$

Let y *be the solution of*

$$(4) \qquad dy(t) = b(y(t))dt + dw(t) \qquad ; \qquad y(0) = x_0.$$

For all v_0, *with probability one*

$$\lim_{\beta \to \infty} x(t) = y(t),$$

uniformly for t *in compact subintervals of* $[0,\infty)$.

<u>Proof.</u> Let κ be the Lipschitz constant of b, so that $|b(x_1)-b(x_2)|\le\kappa|x_1-x_2|$ for all x_1,x_2 in $\mathbb{R}^\ell$. Let

$$t_n = n\frac{1}{2\kappa}$$

for n=0,1,2,.... Consider the equations on $[t_n,t_{n+1}]$. By (2),

$$x(t) = x(t_n) + \int_{t_n}^t v(s)ds, \tag{5}$$

and by (3),

$$v(t) = v(t_n) - \beta\int_{t_n}^t v(s)ds + \beta\int_{t_n}^t b(x(s))ds + \beta[w(t) - w(t_n)], \tag{6}$$

or equivalently,

$$\int_{t_n}^t v(s)ds = \frac{v(t_n)}{\beta} - \frac{v(t)}{\beta} + \int_{t_n}^t b(x(s))ds + w(t) - w(t_n). \tag{7}$$

By (5) and (7),

$$x(t) = x(t_n) + \frac{v(t_n)}{\beta} - \frac{v(t)}{\beta} + \int_{t_n}^t b(x(s))ds + w(t) - w(t_n). \tag{8}$$

By (4),

$$y(t) = y(t_n) + \int_{t_n}^t b(y(s))ds + w(t) - w(t_n), \tag{9}$$

so that by (8) and (9),

$$x(t) - y(t) = x(t_n) - y(t_n) + \frac{v(t_n)}{\beta} - \frac{v(t)}{\beta} + \int_{t_n}^t [b(x(s)) - b(y(s))]ds. \tag{10}$$

The integral in (10) is bounded in absolute value by

$$(t-t_n)\kappa \sup_{t_n\le t\le t_{n+1}} |x(s) - y(s)|,$$

and $(t-t_n)\kappa\le\frac{1}{2}$ for $t_n\le t\le t_{n+1}$, so that

$$(11)\qquad |x(t)-y(t)| \le |x(t_n)-y(t_n)| + 2\sup_{t_n\le s\le t_{n+1}} \left|\frac{v(s)}{\beta}\right| + \frac{1}{2}\sup_{t_n\le s\le t_{n+1}} |x(s)-y(s)|$$

for $t_n \le t \le t_{n+1}$. Since this is true for all such t, we may take the supremum of the left hand side and combine it with the last term on the right hand side. We find

$$(12)\qquad \sup_{t_n\le t\le t_{n+1}} |x(t)-y(t)| \le 2|x(t_n)-y(t_n)| + 4\sup_{t_n\le t\le t_{n+1}} \left|\frac{v(s)}{\beta}\right|.$$

Suppose we can prove that

$$(13)\qquad \sup_{t_n\le t\le t_{n+1}} \left|\frac{v(s)}{\beta}\right| \to 0$$

with probability one as $\beta \to \infty$, for all n. Let

$$\xi_n = \sup_{t_n\le t\le t_{n+1}} |x(t)-y(t)|.$$

Since $x(t_0)-y(t_0)=x_0-x_0=0$, by (12) and (13), $\xi_1 \to 0$ as $\beta \to \infty$. By induction, it follows from (12) and (13) that $\xi_n \to 0$ for all n, which is what we wish to prove. Therefore, we need only prove (13).

If we regard the x(t) as being known, (3) is an inhomogeneous linear equation for v, so that by Theorem 8.2,

$$(14)\qquad v(t) = e^{-\beta(t-t_n)}v(t_n) + \beta\int_{t_n}^{t} e^{-\beta(t-s)}b(x(s))ds + \beta\int_{t_n}^{t} e^{-\beta(t-s)}dw(s).$$

Now consider (8). Since

$$(15)\qquad |b(x(s))| \le |b(x(t_n))| + \kappa|x(s) - x(t_n)|,$$

it follows from (8) that

$$(16)\qquad |x(t) - x(t_n)| \le \left|\frac{v(t_n)}{\beta}\right| + \left|\frac{v(t)}{\beta}\right| + (t-t_n)|b(x(t_n))|$$

$$+ (t-t_n)\kappa \sup_{t_n\le s\le t_{n+1}} |x(s) - x(t_n)| + |w(t) - w(t_n)|.$$

Remembering that $(t-t_n) \le 1/2\kappa$ for $t_n \le t \le t_{n+1}$, we find as before that

$$(17)\qquad \sup_{t_n\le t\le t_{n+1}} |x(t) - x(t_n)| \le 4 \sup_{t_n\le t\le t_{n+1}} \left|\frac{v(t)}{\beta}\right|$$

$$+ \frac{1}{\kappa}|b(x(t_n))| + 2 \sup_{t_n\le t\le t_{n+1}} |w(t) - w(t_n)|.$$

From (15) and (17), we can bound $b(x(s))$ for $t_n \le s \le t_{n+1}$:

$$(18)\qquad \sup_{t_n\le s\le t_{n+1}} |b(x(s))| \le 2|b(x(t_n))|$$

$$+ 4\kappa \sup_{t_n\le t\le t_{n+1}} \left|\frac{v(t)}{\beta}\right| + 2\kappa \sup_{t_n\le t\le t_{n+1}} |w(t) - w(t_n)|.$$

If we apply this to (14) and observe that

$$\beta \int_{t_n}^{t} e^{-\beta(t-s)} ds \le 1,$$

we obtain

$$(19)\qquad \sup_{t_n\le t\le t_{n+1}} |v(t)| \le |v(t_n)| + 2|b(x(t_n))|$$

$$+ 4\kappa \sup_{t_n\le t\le t_{n+1}} \left|\frac{v(t)}{\beta}\right| + 2\kappa \sup_{t_n\le t\le t_{n+1}} |w(t) - w(t_n)|$$

$$+ \sup_{t_n\le t\le t_{n+1}} \left|\beta \int_{t_n}^{t} e^{-\beta(t-s)} dw(s)\right|.$$

Now choose β so large that

$$\frac{4\kappa}{\beta} \leq \frac{1}{2}; \tag{20}$$

i.e., let $\beta \geq 8\kappa$. Then (19) implies that

$$\begin{aligned} \sup_{t_n \leq t \leq t_{n+1}} |v(t)| \leq 2|v(t_n)| + 4|b(x(t_n))| \\ + 4\kappa \sup_{t_n \leq t \leq t_{n+1}} |w(t) - w(t_n)| \\ + 2 \sup_{t_n \leq t \leq t_{n+1}} \left|\beta \int_{t_n}^{t} e^{-\beta(t-s)} dw(s)\right|. \end{aligned} \tag{21}$$

Let

$$\eta_n = \sup_{t_n \leq t \leq t_{n+1}} \left|\frac{v(t)}{\beta}\right|, \tag{22}$$

$$\zeta_n = \left|\frac{b(x(t_n))}{\beta}\right|, \tag{23}$$

$$\begin{aligned} \varepsilon_n = 2 \sup_{t_n \leq t \leq t_{n+1}} \left|\int_{t_n}^{t} e^{-\beta(t-s)} dw(s)\right| \\ + \frac{4\kappa}{\beta} \sup_{t_n \leq t \leq t_{n+1}} |w(t) - w(t_n)|. \end{aligned} \tag{24}$$

Recall that our task is to show that $\eta_n \to 0$ with probability one for all n as $\beta \to \infty$. Suppose we can show that

$$\varepsilon_n \to 0 \tag{25}$$

with probability one for all n as $\beta \to \infty$. By (21),

$$\eta_n \leq 2\eta_{n-1} + 4\zeta_n + \varepsilon_n \tag{26}$$

where $\eta_{-1}=|v_0|/\beta$, and by (18) for n-1 and (20),

(27) $$\zeta_n \leq 2\zeta_{n-1} + \tfrac{1}{2}\eta_{n-1} + \tfrac{1}{2}\varepsilon_n.$$

Now $\zeta_0=|b(x_0)|/\beta \to 0$ and $\eta_{-1}=|v_0|/\beta \to 0$, $\zeta_1 \to 0$ by (27) and (25), and consequently $\eta_1 \to 0$. By induction, $\zeta_n \to 0$ and $\eta_n \to 0$ for all n. Therefore, we need only prove (25).

It is clear that the second term on the right hand side of (24) converges to 0 with probability one as $\beta \to \infty$, since w is continuous with probability one. Let

$$z(t) = \begin{cases} w(t) - w(t_n), & t \geq t_n \\ 0 & , \quad t < t_n. \end{cases}$$

Then

$$\int_{t_n}^{t} e^{-\beta(t-s)} dw(s) = \int_{-\infty}^{t} e^{-\beta(t-s)} dz(s)$$

$$= -\beta\int_{-\infty}^{t} e^{-\beta(t-s)} z(s)ds + z(t).$$

This converges to 0 uniformly for $t_n \leq t \leq t_{n+1}$ with probability one, since z is continuous with probability one. Therefore (25) holds. QED.

A possible physical objection to the theorem is that the initial velocity v_0 should not be held fixed as β varies but should have a Maxwellian distribution (Gaussian with mean 0 and variance $D\beta$). Let v_{00} have a Maxwellian distribution for a fixed value $\beta=\beta_0$. Then $v_0=(\beta/\beta_0)^{\frac{1}{2}}v_{00}$ has a Maxwellian distribution for all β. Since it is still true that $v_0/\beta \to 0$ as $\beta \to \infty$, the theorem remains true with a Maxwellian initial velocity.

Theorem 10.1 has a corollary which may be expressed purely in the language of partial differential equations:

Pseudotheorem 10.2. *Let* $b: \mathbb{R}^\ell \to \mathbb{R}^\ell$ *satisfy a global Lipschitz condition, and let* D *and* β *be strictly positive constants. Let* f_0 *be a bounded continuous function on* $\mathbb{R}^\ell$. *Let* f *on* $[0,\infty)\times\mathbb{R}^\ell$ *be the bounded solution of*

$$(28) \qquad \frac{\partial}{\partial t} f(t,x) = (D\Delta_x + b(x)\cdot\nabla_x)f(t,x); \qquad f(0,x) = f_0(x).$$

Let g_β *on* $[0,\infty)\times\mathbb{R}^\ell\times\mathbb{R}^\ell$ *be the bounded solution of*

$$(29) \qquad \frac{\partial}{\partial t} g_\beta(t,x,v) = (\beta^2 D\Delta_v + v\cdot\nabla_x + \beta(b(x)-v)\cdot\nabla_v)g_\beta(t,x,v);$$

$$g_\beta(0,x,v) = f_0(x).$$

Then for all $t, x,$ *and* v,

$$(30) \qquad \lim_{\beta\to\infty} g_\beta(t,x,v) = f(t,x).$$

To prove this, notice that $f(t,x_0)=Ef_0(y(t))$ and $g_\beta(t,x_0,v_0)=Ef_0(x(t))$, since (28) and (29) are the backward Kolmogoroff equations of the two processes. The result follows from Theorem 10.1 and the Lebesgue dominated convergence theorem.

There is nothing wrong with this proof - only the formulation of the result is at fault. Equation (28) is a parabolic equation with smooth coefficients, and it is a classical result that it has a unique bounded solution. However, (29) is not parabolic (it is of first order in x), so we do not know that it has a unique bounded solution. One way around this problem would be to let $g_{\beta,\varepsilon}$ be the unique bounded solution of (29) with the additional operator $\varepsilon\Delta_x$ on the right hand side and to prove that $g_{\beta,\varepsilon}(t,x_0,v_0) \to g_\beta(t,x_0,v_0)=Ef_0(x(t))$ as $\varepsilon \to 0$. This would give us a characterization of g_β purely in terms of partial differential equations. We shall not do this.

Reference

[26]. Eugen Kappler, Versuche zur Messung der Avogadro-Loschmidtschen Zahl aus der Brownschen Bewegung einer Drehwaage, Annalen der Physik, 11(1931), 233-256.

§11. Kinematics of stochastic motion

We shall investigate the kinematics of motion in which chance plays a rôle (stochastic motion).

Let $x(t)$ be the position of a particle at time t. What does it mean to say that the particle has a velocity $\dot{x}(t)$? It means that if Δt is a very short time interval then

$$x(t+\Delta t) - x(t) = \dot{x}(t)\Delta t + \varepsilon,$$

where ε is a very small percentage error. This is an assumption about actual motion of particles which may or may not be true. Let us be conservative and suppose that it is not necessarily true. ("Conservative" is a useful word for mathematicians. It is used when introducing a hypothesis that a physicist would regard as highly implausible.)

The particle should have some tendency to persist in uniform rectilinear motion for very small intervals of time. Let us use $Dx(t)$ to denote the best prediction we can make, given any relevant information available at time t, of

$$\frac{x(t+\Delta t) - x(t)}{\Delta t}$$

for infinitely small positive Δt.

Let us make this notion precise.

Let I be an interval which is open on the right, let x be an $\mathbb{R}^{\ell}$-valued stochastic process indexed by I, and let $\mathcal{P}_t$ for t in I be an increasing family of σ-algebras such that each $x(t)$ is $\mathcal{P}_t$-measurable. (This implies that $\mathcal{P}_t$ contains the σ-algebra generated by the $x(s)$ with $s \leq t$, $s \in I$. Conversely, this family of σ-algebras satisfies the hypotheses.) We shall have occasion to introduce various regularity assumptions, denoted (R0), (R1), etc.

(R0). Each $x(t)$ is in $\mathcal{L}^1$ and $t \to x(t)$ is continuous from I into $\mathcal{L}^1$.

This is a very weak assumption and by no means implies that the sample functions (trajectories) of the x process are continuous.

(R1). The condition (R0) holds and for each t in I,

$$Dx(t) = \lim_{\Delta t \to 0+} E\{\frac{x(t+\Delta t) - x(t)}{\Delta t} | \mathcal{P}_t\}$$

exists as a limit in $\mathcal{L}^1$, and $t \to Dx(t)$ is continuous from I into $\mathcal{L}^1$.

Here $E\{\ \ | \mathcal{P}_t\}$ denotes the conditional expectation; cf. Doob [15, §6]. The notation $\Delta t \to 0+$ means that Δt tends to 0 through positive values. The random variable $Dx(t)$ is automatically $\mathcal{P}_t$-measurable. It is called the <u>mean forward derivative</u> (or <u>mean forward velocity</u> if $x(t)$ represents position).

As an example of an (R1) process, let $I=(-\infty,\infty)$, let $x(t)$ be the position in the Ornstein-Uhlenbeck process, and let $\mathcal{P}_t$ be the σ-algebra generated by the $x(s)$ with $s \leq t$. Then $Dx(t)=dx(t)/dt=v(t)$. In fact, if $t \to x(t)$ has a continuous strong derivative $dx(t)/dt$ in $\mathcal{L}^1$, then $Dx(t)=dx(t)/dt$. A second example of an (R1) process is a process $x(t)$ of the form discussed in Theorem 8.1, with $I=[0,\infty)$, $x(0)=x_0$, and $\mathcal{P}_t$ the σ-algebra generated by the $x(s)$ with $0 \leq s \leq t$. In this case $Dx(t)=b(x(t))$. The derivative

$dx(t)/dt$ does not exist in this example unless w is identically 0. For a third example, let P^t be a Markovian semigroup on a locally compact Hausdorff space X with infinitesimal generator A, let $I=[0,\infty)$, let the $\xi(t)$ be the X-valued random variables of the Markoff process for some initial measure, and let $\mathcal{P}_t$ be the σ-algebra generated by the $\xi(s)$ with $0\le s\le t$. If f is in the domain of the infinitesimal generator A then $x(t)=f(\xi(t))$ is an (R1) process, and $Df(\xi(t))=Af(\xi(t))$.

<u>Theorem 11.1. Let</u> x <u>be an</u> (R1) <u>process, and let</u> $a\le b$, $a\in I$, $b\in I$. <u>Then</u>

$$E\{x(b)-x(a)\,|\,\mathcal{P}_a\} = E\{\textstyle\int_a^b Dx(s)ds\,|\,\mathcal{P}_a\}. \tag{1}$$

Notice that since $s\to Dx(s)$ is continuous in $\mathcal{L}^1$, the integral exists as a Riemann integral in $\mathcal{L}^1$.

<u>Proof.</u> Let $\varepsilon>0$ and let J be the set of all t in $[a,b]$ such that

$$\|E\{x(s)-x(a)\,|\,\mathcal{P}_a\} - E\{\textstyle\int_a^s Dx(r)dr\,|\,\mathcal{P}_a\}\|_1 \le \varepsilon(s-a) \tag{2}$$

for all $a\le s\le t$, where $\|\ \|_1$ denotes the $\mathcal{L}^1$ norm. Clearly, a is in J, and J is a closed subinterval of $[a,b]$. Let t be the right endpoint of J, and suppose that $t<b$. By the definition of $Dx(t)$, there is a $\delta>0$ such that $t+\delta\le b$ and

$$\|E\{x(t+\Delta t)-x(t)\,|\,\mathcal{P}_t\} - Dx(t)\Delta t\|_1 \le \frac{\varepsilon}{2}\,\Delta t$$

for $0\le\Delta t\le\delta$. Since conditional expectations reduce the $\mathcal{L}^1$ norm and since $\mathcal{P}_t\cap\mathcal{P}_a=\mathcal{P}_a$,

(3) $$\|E\{x(t+\Delta t)-x(t)\mid \mathcal{P}_a\}-E\{Dx(t)\Delta t\mid \mathcal{P}_a\}\|_1 \le \frac{\varepsilon}{2}\Delta t$$

for $0\le\Delta t\le\delta$. By reducing δ if necessary, we find

$$\|Dx(t)\Delta t-\int_t^{t+\Delta t}Dx(s)ds\|_1 \le \frac{\varepsilon}{2}\Delta t$$

for $0\le\Delta t\le\delta$, since $s\to Dx(s)$ is $\mathcal{L}^1$ continuous. Therefore,

(4) $$\|E\{Dx(t)\Delta t\mid \mathcal{P}_a\}-E\{\int_t^{t+\Delta t}Dx(s)ds\mid \mathcal{P}_a\}\|_1 \le \frac{\varepsilon}{2}\Delta t$$

for $0\le\Delta t\le\delta$. From (2) for $s=t$, (3), and (4), it follows that (2) holds for all $t+\Delta t$ with $0\le\Delta t\le\delta$. This contradicts the assumption that t is the right end-point of J, so we must have $t=b$. Since ε is arbitrary, (1) holds. QED.

Theorem 11.1 and its proof remain valid without the assumption that $x(t)$ is $\mathcal{P}_t$-measurable.

Theorem 11.2. *An (R1) process is a martingale if and only if $Dx(t)=0$, $t\in I$. It is a submartingale if and only if $Dx(t)\ge 0$, $t\in I$ and a supermartingale if and only if $Dx(t)\le 0$, $t\in I$.*

We mean, of course, martingale, etc., relative to the $\mathcal{P}_t$. This theorem is an immediate consequence of Theorem 11.1 and the definitions (see Doob [15, p.294]). Note that in the older terminology, "semimartingale" means submartingale and "lower semimartingale" means supermartingale.

Given an (R1) process x, define the random variable $y(a,b)$, for all a and b in I, by

(5) $$x(b)-x(a) = \int_a^b Dx(s)ds + y(a,b).$$

We always have $y(b,a)=-y(a,b)$, $y(a,b)+y(b,c)=y(a,c)$, and $y(a,b)$ is

$\mathcal{B}_{\max(a,b)}$-measurable, for all a,b, and c in I. We call a stochastic process indexed by I×I which has these three properties a <u>difference process.</u> If y is a difference process, we may choose a point a_0 in I, define $y(a_0)$ arbitrarily (say $y(a_0)=0$), and define y(b) for all b in I by $y(b)=y(a_0,b)$. Then $y(a,b)=y(b)-y(a)$ for all a and b in I. The only trouble is that y(b) will not in general be $\mathcal{B}_b$-measurable for $b<a_0$. If I has a left end-point, we may choose a_0 to be it and then y(b) will always be $\mathcal{B}_b$-measurable. By Theorem 11.1, $E\{y(b)-y(a)|\mathcal{B}_a\}=0$ whenever $a\leq b$, so that when a_0 is the left end-point of I, y(b) is a martingale relative to the $\mathcal{B}_b$. In the general case, we call a difference process y(a,b) such that $E\{y(a,b)|\mathcal{B}_a\}=0$ whenever $a\leq b$ a <u>difference martingale</u>. The following is an immediate consequence of Theorem 11.1.

<u>Theorem 11.3. Let</u> x <u>be an</u> (R1) <u>process, and define</u> y <u>by (5). Then</u> y <u>is a difference martingale.</u>

From now on we shall write y(b)-y(a) instead of y(a,b) when y is a difference process.

We introduce another regularity condition, denoted (R2). It is a regularity condition on a difference martingale y. If it holds, we say that y is an (R2) difference martingale, and if in addition y is defined in terms of an (R1) process x by (5) then we say that x is an (R2) process.

(R2). For each a and b in I, y(b)-y(a) is in $\mathcal{L}^2$. For each t in I,

$$(6) \qquad \sigma^2(t) = \lim_{\Delta t \to 0+} E\{\frac{[y(t+\Delta t)-y(t)]^2}{\Delta t}|\mathcal{B}_t\}$$

exists in $\mathcal{L}^1$, and $t \to \sigma^2(t)$ is continuous from I into $\mathcal{L}^1$.

The process y has values in $\mathbb{R}^\ell$. In case $\ell>1$, the expression $[y(t+\Delta t)-y(t)]^2$ means $[y(t+\Delta t)-y(t)]\otimes[y(t+\Delta t)-y(t)]$ and $\sigma^2(t)$ is a matrix of positive type.

Observe that Δt occurs to the first power in (6) while $[y(t+\Delta t)-y(t)]$ occurs to the second power.

Theorem 11.4. Let y be an (R2) difference martingale, and let $a\leq b$, $a\in I$, $b\in I$. Then

(7) $$E\{[y(b)-y(a)]^2 \mid \mathcal{B}_a\} = E\{\int_a^b \sigma^2(s)ds \mid \mathcal{B}_a\}.$$

The proof is so similar to the proof of Theorem 11.1 that it will be omitted.

Next we shall discuss the Itô-Doob stochastic integral, which is a generalization of the Wiener integral. The new feature is that the integrand is a random variable depending on the past history of the process.

Let y be an (R2) difference martingale. Let $\mathcal{H}_0$ be the set of functions of the form

(8) $$f = \sum_{i=1}^{n} \varphi_{[a_i,b_i]},$$

where the intervals $[a_i,b_i]$ are non-overlapping intervals in I and each f_i is a real-valued $\mathcal{B}_{a_i}$-measurable random variable in $\mathcal{L}^2$. (The symbol φ denotes characteristic function.) Thus each f in $\mathcal{H}_0$ is a stochastic process indexed by I. For each f given by (8) we define the stochastic integral

$$\int f(t)dy(t) = \sum_{i=1}^{n} f_i[y(b_i)-y(a_i)].$$

This is a random variable.

For f in $\mathcal{H}_0$,

(9) $$E[\int f(t)dy(t)]^2 = \sum_{i,j=1}^{n} Ef_i[y(b_i) - y(a_i)]f_j[y(b_j) - y(a_j)].$$

If i<j then $f_i[y(b_i)-y(a_i)]f_j$ is $\mathcal{B}_{a_j}$-measurable, and $E\{y(b_j)-y(a_j)|\mathcal{B}_{a_j}\}=0$ since y is a difference martingale. Therefore the terms with i<j in (9) are 0, and similarly for the terms with i>j. The terms with i=j are

$$Ef_i^2E\{\int_{a_i}^{b_i}\sigma^2(s)ds|\mathcal{B}_{a_i}\} = \int_{a_i}^{b_i} Ef_i^2\sigma^2(s)ds$$

by (7). Therefore

$$E[\int f(t)dy(t)]^2 = \int_I Ef^2(t)\sigma^2(t)dt.$$

This is a matrix of positive type. If we give $\mathcal{H}_0$ the norm

(10) $$\|f\|^2 = \text{tr} \int_I Ef^2(t)\sigma^2(t)dt$$

then $\mathcal{H}_0$ is a pre-Hilbert space, and the mapping $f \to \int f(t)dy(t)$ is isometric from $\mathcal{H}_0$ into the real Hilbert space of square-integrable $\mathbb{R}^\ell$-valued random variables, which will be denoted by $\mathcal{L}^2_{\mathbb{R}^\ell}$.

Let $\mathcal{H}$ be the completion of $\mathcal{H}_0$. The mapping $f \to \int f(t)dy(t)$ extends uniquely to be unitary from $\mathcal{H}$ into $\mathcal{L}^2_{\mathbb{R}^\ell}$. Our problem now is to describe $\mathcal{H}$ in concrete terms.

Let $\sigma(t)$ be the positive square root of $\sigma^2(t)$. If f is in $\mathcal{H}_0$ then $f\sigma$ is square-integrable. If f_j is a Cauchy sequence in $\mathcal{H}_0$ then $f_j\sigma$ converges in the $\mathcal{L}^2$ norm, so that a subsequence, again denoted by f_j, converges a.e. to a square-integrable matrix-valued function g on I. Therefore f_j converges for a.e. t such that $\sigma(t) \neq 0$. Let us define $f(t) = \lim f_j(t)$ when the limit exists and define f arbitrarily to be 0 when the limit does not exist. Then $\|f_j - f\| \to 0$, and $f(t)\sigma(t)$ is a $\mathcal{B}_t$-measurable square-integrable random variable for a.e. t. By definition of strong measurability [14, §5], $f\sigma$ is strongly measurable. Let $\mathcal{K}$ be the set of all functions f, defined a.e. on I, such that $f\sigma$ is a strongly measurable square-integrable function with f(t) $\mathcal{B}_t$-measurable for a.e. t. We have seen that every element of $\mathcal{H}$ can be identified with an element of $\mathcal{K}$, uniquely defined except on sets of measure 0.

Conversely, let f be in $\mathcal{K}$. We wish to show that it can be approximated arbitrarily closely in the norm (10) by an element of $\mathcal{H}_0$. Firstly, f can be approximated arbitrarily closely by an element of $\mathcal{K}$ with support contained in a compact interval I_0 in I, so we may as well assume that f has support in I_0. Let

$$f_k(t) = \begin{cases} k \;, & f(t) > k \\ f(t), & |f(t)| \leq k \\ -k \;, & f(t) < -k. \end{cases}$$

Then $\|f_k - f\| \to 0$ as $k \to \infty$, so we may as well assume that f is uniformly bounded (and consequently has uniformly bounded $\mathcal{L}^2$ norm). Divide I_0 into n equal parts, and let f_n be the function which on each subinterval is the average (Bochner integral [14, §5]) of f on the preceding subinterval (and let f_n be 0 on the first subinterval). Then f_n is in $\mathcal{H}_0$ and $\|f_n - f\| \to 0$.

With the usual identification of functions equal a.e., we

may identify $\mathcal{H}$ and $\mathcal{K}$. We have proved the following theorem.

Theorem 11.5. *Let $\mathcal{H}$ be the Hilbert space of functions f defined a.e. on I such that $f\sigma$ is strongly measurable and square-integrable and such that f(t) is $\mathcal{P}_t$-measurable for a.e. t, with the norm (10). There is a unique unitary mapping $f \to \int f(y)dy(t)$ from $\mathcal{H}$ into $\mathcal{L}^2_{\mathbb{R}^\ell}$ such that if $f=f_0\varphi_{[a,b]}$ where $a\leq b$, $a\epsilon I$, $b\epsilon I$, $f_0 \epsilon \mathcal{L}^2$, f_0 $\mathcal{P}_a$-measurable, then*

$$\int f(t)dy(t) = f_0[y(b) - y(a)].$$

We now introduce our last regularity hypothesis.

(R3). For a.e. t in I, $\det \sigma^2(t) > 0$ a.e.

An (R2) difference martingale for which this holds will be called an (R3) difference martingale. An (R2) process x for which the associated difference martingale y satisfies this will be called an (R3) process.

Let $\sigma^{-1}(t)=\sigma(t)^{-1}$, where $\sigma(t)$ is the positive square root of $\sigma^2(t)$.

Theorem 11.6. *Let x be an (R3) process. Then there is a difference martingale w such that*

$$E\{[w(b) - w(a)]^2 | \mathcal{P}_a\} = b-a$$

and

$$x(b) - x(a) = \int_a^b Dx(s)ds + \int_a^b \sigma(s)dw(s)$$

whenever $a\leq b$, $a\epsilon I$, $b\epsilon I$.

Proof. Let

$$w(a,b) = \int_a^b \sigma^{-1}(s)dy(s).$$

This is well-defined, since each component of $\sigma^{-1}\varphi_{[a,b]}$ is in $\mathcal{H}$. If f is in $\mathcal{H}_0$, a simple computation shows that $\int_a^b f(s)dy(s)$ is a difference martingale, so the same is true if f is in $\mathcal{H}$ or if each $f\varphi_{[a,b]}$ is in $\mathcal{H}$. Therefore, w is a difference martingale, and we will write w(b)-w(a) for w(a,b).

If f is in $\mathcal{H}_0$, given by (8), and if f(t)=0 for all t<a, then

$$\begin{aligned}
&E\{[\int f(t)dy(t)]^2 | \mathcal{B}_a\} = \\
&E\{\sum_i f_i^2[y(b_i) - y(a_i)]^2 | \mathcal{B}_a\} = \\
&E\{E\{\sum_i f_i^2 \int_{a_i}^{b_i} \sigma^2(s)ds | \mathcal{B}_{a_i}\} | \mathcal{B}_a\} = \\
&E\{\sum_i f_i^2 \int_{a_i}^{b_i} \sigma^2(s)ds | \mathcal{B}_a\} = \\
&E\{\int f^2(s)\sigma^2(s)ds | \mathcal{B}_a\}.
\end{aligned}$$

By continuity,

$$\begin{aligned}
&E\{[\int_a^b f(t)dy(t)]^2 | \mathcal{B}_a\} = \\
&E\{\int_a^b f^2(s)\sigma^2(s)ds | \mathcal{B}_a\}
\end{aligned}$$

for all f in $\mathcal{H}$. If we apply this to the components of $\sigma^{-1}\varphi_{[a,b]}$ we find

$$\begin{aligned}
&E\{[w(b) - w(a)]^2 | \mathcal{B}_a\} = \\
&E\{\int_a^b \sigma^{-1}(s)\sigma^2(s)\sigma^{-1}(s)ds | \mathcal{B}_a\} = \\
&b-a,
\end{aligned}$$

whenever $a \leq b$, $a \in I$, $b \in I$. Consequently, w is an (R2) (in fact, (R3)) difference martingale, and the corresponding σ^2 is identically 1.

Therefore we may construct stochastic integrals with respect to w.

Formally, $dw(t)=\sigma^{-1}(t)dy(t)$, so that $dy(t)=\sigma(t)dw(t)$. Let us prove that, in fact,

$$y(b) - y(a) = \int_a^b \sigma(s)dw(s).$$

A simple calculation shows that if f is in $\mathcal{H}_0$ then

$$\int f(s)dw(s) = \int f(s)\sigma^{-1}(s)dy(s).$$

Consequently, the same holds for any f in $\mathcal{H}$. Therefore,

$$\int_a^b \sigma(s)dw(s) = \int_a^b \sigma(s)\sigma^{-1}(s)dy(s)$$
$$= y(b) - y(a) = y(a,b).$$

The theorem follows from the definition (5) of y. QED.

It is possible that the theorem remains true without the regularity assumption (R3) provided that one is allowed to enlarge the underlying probability space and the σ-algebras $\mathcal{B}_t$.

A fundamental aspect of motion has been neglected in the discussion so far; to wit, the continuity of motion. We shall assume from now on that (with probability one) the sample functions of x are continuous. By (5), this means that the same functions of y are continuous. (Use ω to denote a point in the underlying probability space. We may choose a version $Dx(s,\omega)$ of the stochastic process Dx which is jointly measurable in s and ω, since $s \to Dx(s)$ is continuous in $\mathcal{L}^1$ [15, §6, p.60 ff]. Then $\int_a^b Dx(s,\omega)ds$ is in fact absolutely continuous as b varies, so that $y(b)-y(a)$ has continuous sample paths as b varies.) Next we show (following Doob [15, p.446]) that this implies that ω has continuous sample functions.

<u>Theorem 11.7.</u> <u>Let</u> y <u>be an</u> (R2) <u>difference martingale whose sample functions are continuous with probability one, and let</u> f <u>be in</u> $\mathcal{H}$. <u>Let</u>

$$z(b) - z(a) = \int_a^b f(s)dy(s).$$

<u>Then</u> z <u>is a difference martingale whose sample paths are continuous with probability one.</u>

<u>Proof.</u> If f is in $\mathcal{H}_0$, this is evident. If f is in $\mathcal{H}$, let f_n be in $\mathcal{H}_0$ with $\|f-f_n\| \leq 1/n^2$, where the norm is given by (10). Let

$$z_n(b) - z_n(a) = \int_a^b f_n(s)dy(s).$$

Then $z-z_n$ is a difference martingale. (We already observed in the proof of Theorem 11.6 that z is a difference martingale - only the continuity of sample functions is at issue.)

By the Kolmogorov inequality for martingales (Doob [15, p.105]), if S is any finite subset of [a,b],

$$\Pr\{\sup_{s \in S} |z(s) - z_n(s)| > \frac{1}{n}\} \leq \frac{1}{n^4} \cdot n^2 = \frac{1}{n^2}.$$

Since S is arbitrary, we have

$$\Pr\{\sup_{a \leq s \leq b} |z(s) - z_n(s)| > \frac{1}{n}\} \leq \frac{1}{n^2}.$$

(This requires a word concerning interpretation, since the supremum is over an uncountable set. We may either assume that $z-z_n$ is separable in the sense of Doob or take the product space representation as in [25, §9] of the pair z, z_n.) By the Borel-Cantelli lemma, z_n converges uniformly on [a,b] to z. QED.

Notice that we only need f to be locally in $\mathcal{H}$; i.e., we only need $f\varphi_{[a,b]}$ to be in $\mathcal{H}$ for [a,b] any compact subinterval of I. In particular, if y is an (R3) process the above result applies to each component of σ^{-1}, so that w has continuous sample paths if y does.

Now we shall study the difference martingale w (with σ^2 identically 1) under the assumption that w has continuous sample paths.

<u>Theorem 11.8.</u> <u>Let</u> w <u>be a difference martingale in</u> $\mathbb{R}^\ell$ <u>satisfying</u>

$$E\{[w(b)-w(a)]^2 \mid \mathcal{P}_a\} = b-a$$

<u>whenever</u> $a\leq b$, $a\in I$, $b\in I$, <u>and having continous sample paths with probability one. Then</u> w <u>is a Wiener process.</u>

<u>Proof.</u> We need only show that the w(b)-w(a) are Gaussian. There is no loss of generality in assuming that a=0 and b=1. First we assume that $\ell=1$.

Let Δt be the reciprocal of a strictly positive integer and let $\Delta w(t)=w(t+\Delta t)-w(t)$. Then

$$[w(1)-w(0)]^n = \Sigma\, \Delta w(t_1)\ldots\Delta w(t_n),$$

where the sum is over all $t_1,\ldots,t_n$ ranging over $0,\Delta t,2\Delta t,\ldots,1-\Delta t$.

We write the sum as $\Sigma=\Sigma'+\Sigma''$, where Σ' is the sum of all terms in which no three of the t_i are equal.

Let B(K) be the set such that $|w(1)-w(0)|\leq K$. Then

$$\lim_{K\to\infty} \Pr(B(K)) = 1.$$

Let $\Gamma(\varepsilon,\delta)$ be the set such that $|w(t)-w(s)|\le\varepsilon$ whenever $|t-s|\le\delta$, for $0\le t,s\le 1$. Since w has continuous sample paths with probability one,

$$\lim_{\delta\to 0} \Pr(\Gamma(\varepsilon,\delta)) = 1$$

for each $\varepsilon>0$.

Let $\alpha>0$. Choose $K\ge 1$ so that $\Pr(B(K))\ge 1-\alpha$. Given n, choose ε so small that $nK^n\varepsilon\le\alpha$ and then choose δ so small that $\Pr(\Gamma(\varepsilon,\delta))\ge 1-\alpha$. Now the sum Σ'' may be written

$$\Sigma'' = \Sigma_0'' + \Sigma_1'' + \dots + \Sigma_{n-3}'',$$

where Σ_ν'' means that exactly ν of the t_i are distinct and some three of the t_i are equal. Then Σ_ν'' has a factor $[w(1)-w(0)]^\nu$ times a sum of terms in which all t_i which occur, occur at least twice, and in which at least one t_i occurs at least thrice. Therefore, if $\Delta t\le\delta$,

$$\left|\int_{\Gamma(\varepsilon,\delta)\cap B(K)} \Sigma_\nu'' \, d\Pr\right| \le K^\nu\varepsilon\int \Sigma\Delta w(t_1)^2\dots\Delta w(t_j)^2 d\Pr \le K^\nu\varepsilon,$$

where the $t_1,\dots,t_j$ are distinct. Therefore

$$\left|\int_{\Gamma(\varepsilon,\delta)\cap B(K)} \Sigma'' \, d\Pr\right| \le nK^n\varepsilon \le \alpha.$$

Those terms in Σ' in which one or more of the t_i occurs only once have expectation 0, so

$$\int \Sigma' d\Pr = \mu_n,$$

where $\mu_n=0$ if n is odd and $\mu_n=(n-1)(n-3)\dots 5\cdot 3\cdot 1$ if n is even, since this is the number of ways of dividing n objects into distinct pairs.

Consequently, the integral of $[w(1)-w(0)]^n$ over a set of arbitrarily large measure is arbitrarily close to μ_n. If n is even, the integrand $[w(1)-w(0)]^n$ is positive, so this shows that $[w(1)-w(0)]^n$ is integrable for all even n and hence for all n. Therefore,

$$E[w(1)-w(0)]^n = \mu_n$$

for all n. But the μ_n are the moments of the Gaussian measure with mean 0 and variance 1, and they increase slowly enough for uniqueness to hold in the moment problem. In fact,

$$Ee^{i\lambda[w(1)-w(0)]} = E\sum_{n=0}^{\infty}\frac{(i\lambda)^n}{n!}[w(1)-w(0)]^n =$$

$$= \sum_{n=0}^{\infty}\frac{(i\lambda)^n}{n!}\mu_n = e^{-\frac{\lambda^2}{2}},$$

so that $w(1)-w(0)$ is Gaussian.

The proof for $\ell>1$ goes the same way, except that all products are tensor products. For example, $(n-1)(n-3)\ldots3\cdot1$ is replaced by

$$(n-1)\delta_{i_1i_2}(n-3)\delta_{i_3i_4}\ldots3\delta_{i_{n-3}i_{n-2}}1\delta_{i_{n-1}i_n}.$$

QED.

We summarize the results obtained so far in the following theorem.

Theorem 11.9. Let I be an interval open on the right, $\mathcal{B}_t$ (for $t\in I$) an increasing family of σ-algebras of measurable sets on a probability space, x a stochastic process on $\mathbb{R}^\ell$ having continuous sample paths with probability one, such that each $x(t)$ is $\mathcal{B}_t$-measurable and such that

$$Dx(t) = \lim_{\Delta t \to 0+} E\{\frac{x(t+\Delta t)-x(t)}{\Delta t} | \mathcal{P}_t\}$$

<u>and</u>

$$\sigma^2(t) = \lim_{\Delta t \to 0+} E\{\frac{[x(t+\Delta t)-x(t)]^2}{\Delta t} | \mathcal{P}_t\}$$

<u>exist in</u> $\mathcal{L}^1$ <u>and are</u> $\mathcal{L}^1$ <u>continuous in</u> t, <u>and such that</u> $\sigma^2(t)$ <u>is a.e. invertible for a.e.</u> t. <u>Then there is a Wiener process</u> w <u>on</u> $\mathbb{R}^\ell$ <u>such that each</u> $w(t)-w(s)$ <u>is</u> $\mathcal{P}_{\max(t,s)}$<u>-measurable, and</u>

$$x(b) - x(a) = \int_a^b Dx(s)ds + \int_a^b \sigma(s)dw(s)$$

<u>for all</u> a <u>and</u> b <u>in</u> I.

* * * * *

So far we have been adopting the standard viewpoint of the theory of stochastic processes, that the past is known and that the future develops from the past according to certain probabilistic laws. Nature, however, operates on a different scheme in which the past and the future are on an equal footing. Consequently it is important to give a treatment of stochastic motion in which a complete symmetry between past and future is maintained.

Let I be an open interval, let x be an $\mathbb{R}^\ell$-valued stochastic process indexed by I, let $\mathcal{P}_t$ for t in I be an increasing family of σ-algebras such that each x(t) is $\mathcal{P}_t$-measurable, and let $\mathcal{F}_t$ be a decreasing family of σ-algebras such that each x(t) is $\mathcal{F}_t$-measurable. ($\mathcal{P}_t$ represents the past, $\mathcal{F}_t$ the future.) The following regularity conditions make the conditions (R1), (R2), and (R3) symmetric with respect to past and future. The condition (R0) is already symmetric.

(S1). The condition (R1) holds and, for each t in I,

$$D_*x(t) = \lim_{\Delta t \to 0+} E\{\frac{x(t)-x(t-\Delta t)}{\Delta t} | \tilde{\mathcal{F}}_t\}$$

exists as a limit in $\mathcal{L}^1$, and $t \to D_*x(t)$ is continuous from I into $\mathcal{L}^1$.

Notice that the notation is chosen so that if $t \to x(t)$ is strongly differentiable in $\mathcal{L}^1$ then $Dx(t)=D_*x(t)=dx(t)/dt$. The random variable $D_*x(t)$ is called the <u>mean backward derivative</u> or <u>mean backward velocity</u>, and is in general different from $Dx(t)$.

We define $y_*(a,b)=y_*(b)-y_*(a)$ by

$$x(b) - x(a) = \int_a^b D_*x(s)ds + y_*(b) - y_*(a).$$

It is a difference martingale relative to the $\tilde{\mathcal{F}}_t$ with the direction of time reversed.

(S2). The conditions (R2) and (S1) hold and, for each t in I,

$$\sigma_*^2(t) = \lim_{\Delta t \to 0+} E\{\frac{[y(t)-y(t-\Delta t)]^2}{\Delta t} | \tilde{\mathcal{F}}_t\}$$

exists as a limit in $\mathcal{L}^1$ and $t \to \sigma_*^2(t)$ is continuous from I into $\mathcal{L}^1$.

(S3). The conditions (R3) and (S2) hold and $\det \sigma_*{}^2(t)>0$ a.e. for a.e. t.

We obtain theorems analogous to the preceding ones. In particular, if $a\leq b$, $a\epsilon I$, $b\epsilon I$, then for an (S1) process

(11) $$E\{x(b) - x(a) \mid \mathcal{F}_b\} = E\{\int_a^b D_* x(s)ds \mid \mathcal{F}_b\},$$

and for an (S2) process

(12) $$E\{[y_*(b) - y_*(a)]^2 \mid \mathcal{F}_b\} = E\{\int_a^b \sigma_*^2(s)ds \mid \mathcal{F}_b\}.$$

Theorem 11.10. *Let x be an (S1) process. Then*

(13) $$EDx(t) = ED_*x(t)$$

for all t in I. Let x be an (S2) process. Then

(14) $$E\sigma^2(t) = E\sigma_*^2(t)$$

for all t in I.

Proof. By Theorem 11.1 and (11), if we take absolute expectations we find

$$E[x(b) - x(a)] = E\int_a^b Dx(s)ds = E\int_a^b D_*x(s)ds$$

for all a and b in I. Since $s \to Dx(s)$ and $s \to D_*x(s)$ are continuous in $\mathcal{L}^1$, (13) holds. Similarly, (14) follows from Theorem 11.4 and (12). QED.

Theorem 11.11. *Let x be an (S1) process. Then x is a constant (i.e., x(t) is the same random variable for all t) if and only if* $Dx = D_*x = 0$.

Proof. The only if part of the theorem is trivial. Suppose that $Dx = D_*x = 0$. By Theorem 11.2, x is a martingale and a martingale with the direction of time reversed. Let $t_1 \neq t_2$, $x_1 = x(t_1)$, $x_2 = x(t_2)$. Then x_1 and x_2 are in $\mathcal{L}^1$ and $E\{x_1|x_2\} = x_2$, $E\{x_2|x_1\} = x_1$. We wish to show that $x_1=x_2$ (a.e., of course).

If x_1 and x_2 are in $\mathcal{L}^2$ (as they are if x is an (S2) process) there is a trivial proof, as follows. We have

$$E\{(x_2 - x_1)^2|x_1\} = E\{x_2^2 - 2x_2x_1 + x_1^2|x_1\} = E\{x_2^2|x_1\} - x_1^2,$$

so that if we take absolute expectations we find

$$E(x_2 - x_1)^2 = Ex_2^2 - Ex_1^2.$$

The same result holds with x_1 and x_2 interchanged. Thus $E(x_2-x_1)^2=0$, $x_2=x_1$ a.e.

G. A. Hunt showed me the following proof for the general case (x_1,x_2 in $\mathcal{L}^1$).

Let μ be the distribution of x_1,x_2 in the plane. We may take x_1 and x_2 to be the coordinate functions. Then there is a conditional probability distribution $p(x_1,\cdot)$ such that if ν is the distribution of x_1 and f is a positive Baire function on $\mathbb{R}^2$,

$$\int f(x_1,x_2)d\mu(x_1,x_2) = \int\int f(x_1,x_2)p(x_1,dx_2)d\nu(x_1).$$

(See Doob [15, §6, pp.26-34].) Then

$$E\{\varphi(x_2)|x_1\} = \int\varphi(x_2)p(x_1,dx_2) \quad \text{a.e.}[\nu]$$

provided $\varphi(x_2)$ is in $\mathcal{L}^1$. Take φ to be strictly convex with $|\varphi(\xi)|\leq|\xi|$ for all real ξ (so that $\varphi(x_2)$ is in $\mathcal{L}^1$). Then, for each x_1, since φ is strictly convex, Jensen's inequality gives

$$\varphi(\int x_2 p(x_1,dx_2)) < \int\varphi(x_2)p(x_1,dx_2)$$

unless $\varphi(x_2)=\int\varphi(x_2)p(x_1,dx_2)$ a.e.$[p(x_1,\cdot)]$. But

$$\int x_2 p(x_1,dx_2) = x_1 \text{ a.e.}[\nu],$$

so, unless $x_2=x_1$ a.e.$[\nu]$,

$$\varphi(x_1) < \int\varphi(x_2)p(x_1,dx_2).$$

If we take absolute expectations, we find $E\varphi(x_1)<E\varphi(x_2)$ unless $x_2=x_1$ a.e. The same argument gives the reverse inequality, so $x_2=x_1$ a.e. QED.

Theorem 11.12. *Let x and y be (S1) processes with respect to the same families of σ-algebras $\mathcal{B}_t$ and $\tilde{\mathcal{F}}_t$, and suppose that $x(t)$, $y(t)$, $Dx(t)$, $Dy(t)$, $D_*x(t)$, and $D_*y(t)$ all lie in $\mathcal{L}^2$ and are continuous functions of t in $\mathcal{L}^2$. Then*

$$\frac{d}{dt} Ex(t)y(t) = EDx(t)\cdot y(t) + Ex(t)D_*y(t).$$

Proof. We need to show, for a and b in I, that

$$E[x(b)y(b) - x(a)y(a)] = \int_a^b E[Dx(t)\cdot y(t) + x(t)D_*y(t)]dt.$$

(Notice that the integrand is continuous.) Divide $[a,b]$ into n equal parts: $t_j=a+j(b-a)/n$ for $j=0,\ldots,n$. Then

$$E[x(b)y(b) - x(a)y(a)] = \lim_{n\to\infty} \sum_{j=1}^{n-1} E[x(t_{j+1})y(t_j) - x(t_j)y(t_{j-1})] =$$

$$\lim_{n\to\infty} \sum_{j=1}^{n-1} E[(x(t_{j+1}) - x(t_j)) \frac{y(t_j)+y(t_{j-1})}{2} +$$

$$\frac{x(t_{j+1})+x(t_j)}{2} (y(t_j) - y(t_{j-1}))] =$$

$$\lim_{n\to\infty} \sum_{j=1}^{n-1} E[Dx(t_j)\cdot y(t_j) + x(t_j)D_*y(t_j)] \frac{b-a}{n} =$$

$$\int_a^b E[Dx(t)\cdot y(t) + x(t)D_*y(t)]dt.$$ QED.

Now let us assume that the past $\mathcal{P}_t$ and the future $\mathcal{F}_t$ are conditionally independent given the present $\mathcal{P}_t \cap \mathcal{F}_t$. That is, we assume that if f is any $\mathcal{F}_t$-measurable function in $\mathcal{L}^1$ then $E\{f|\mathcal{P}_t\} = E\{f|\mathcal{P}_t \cap \mathcal{F}_t\}$, and if f is any $\mathcal{P}_t$-measurable function in $\mathcal{L}^1$ then $E\{f|\mathcal{F}_t\} = E\{f|\mathcal{P}_t \cap \mathcal{F}_t\}$. If x is a Markoff process and $\mathcal{P}_t$ is generated by the x(s) with $s \leq t$, $\mathcal{F}_t$ by the x(s) with $s \geq t$, this is certainly the case. However, the assumption is much weaker. It applies, for example, to the position x(t) of the Ornstein-Uhlenbeck process. The reason is that the present $\mathcal{P}_t \cap \mathcal{F}_t$ may not be generated by x(t); for example, in the Ornstein-Uhlenbeck case $v(t)=dx(t)/dt$ is also $\mathcal{P}_t \cap \mathcal{F}_t$-measurable.

With the above assumption on the $\mathcal{P}_t$ and $\mathcal{F}_t$, if x is an (S1) process then $Dx(t)$ and $D_*x(t)$ are $\mathcal{P}_t \cap \mathcal{F}_t$-measurable, and we may form $DD_*x(t)$ and $D_*Dx(t)$ if they exist. Assuming they exist, we define

$$a(t) = \tfrac{1}{2}DD_*x(t) + \tfrac{1}{2}D_*Dx(t) \tag{15}$$

and call it the <u>mean second derivative</u> or <u>mean acceleration</u>.

If x is a sufficiently smooth function of t then $a(t)=d^2x(t)/dt^2$. This is also true of other possible candidates for the title of mean acceleration, such as $DD_*x(t)$, $D_*Dx(t)$, $DDx(t)$, $D_*D_*(t)$, and $\frac{1}{2}DDx(t)+\frac{1}{2}D_*D_*x(t)$. Of these the first four distinguish between the two choices of direction for the time axes, and so may be discarded. To discuss the fifth possibility, consider the Gaussian Markoff process x(t) satisfying

$$dx(t) = -\omega x(t)dt + dw(t),$$

where w is a Wiener process, in equilibrium (that is, with the invariant Gaussian measure as initial measure). Then

$$Dx(t) = -\omega x(t),$$

$$D_*x(t) = \omega x(t),$$

$$a(t) = -\omega^2 x(t),$$

but

$$\frac{1}{2}DDx(t) + \frac{1}{2}D_*D_*x(t) = \omega^2 x(t).$$

This process is familiar to us: it is the position in the Smoluchowski description of the highly overdamped harmonic oscillator (or the velocity of a free particle in the Ornstein-Uhlenbeck theory). The characteristic feature of this process is its constant tendency to go towards the origin, no matter which direction of time is taken. Our definition of mean acceleration, which gives $a(t)=-\omega^2 x(t)$, is kinematically the appropriate definition.

Reference

The stochastic integral was invented by Itô,

[27]. Kiyosi Itô, On Stochastic Differential Equations, Memoirs of the American Mathematical Society, Number 4(1951).

Doob gave a treatment based on martingales [15, §6, pp.436-451]. Our discussion of stochastic integrals, as well as most of the other material of this section, is based on Doob's book.

§12. Dynamics of stochastic motion

The fundamental law of non-relativistic dynamics is Newton's law F=ma: the force on a particle is the product of the particle's mass and the acceleration of the particle. This law is, of course, nothing but the definition of force. Most definitions are trivial - others are profound. Feynman [28] has analyzed the characteristics which make Newton's definition profound:

"It implies that if we study the mass times the acceleration and call the product the force, i.e., if we study the characteristics of force as a program of interest, then we shall find that forces have some simplicity; the law is a good program for analyzing nature, it is a suggestion that the forces will be simple."

Now suppose that x is a stochastic process representing the motion of a particle of mass m. Leaving unanalyzed the dynamical mechanism causing the random fluctuations, we may ask how to express the fact that there is an external force F acting on the particle. We do this simply by setting

$$F = ma$$

where a is the mean acceleration (§11).

For example, suppose that x is the position in the Ornstein-Uhlenbeck theory of Brownian motion, and suppose that the external force is $F = -\text{grad } V$ where $\exp(-VD/m\beta)$ is integrable. In equilibrium, the particle has probability density a normalization constant times $\exp(-VD/m\beta)$ and satisfies

$$dx(t) = v(t)dt,$$

$$dv(t) = -\beta v(t)dt + K(x(t))dt + dB(t),$$

where K=F/m=-grad V/m, and B has variance parameter $2\beta^2 D$. Then

$$\begin{aligned} Dx(t) &= D_* x(t) = v(t), \\ Dv(t) &= -\beta v(t) + K(x(t)), \\ D_* v(t) &= \beta v(t) + K(x(t)), \\ a(t) &= K(x(t)). \end{aligned}$$

Therefore the law F=ma holds.

Reference

[28]. Richard P. Feynman, Robert B. Leighton, and Matthew Sands, The Feynman Lectures on Physics, Addison-Wesley, Reading, Massachusetts, 1963.

§13. Kinematics of Markoffian motion

At this point I will cease making regularity assumptions explicit. Whenever we take the derivative of a function, the function is assumed to be differentiable. Whenever we take D of a stochastic process, it is assumed to exist. Whenever we consider the probability density of a random variable, it is assumed to exist. I do this not out of laziness but out of ignorance. The problem of finding convenient regularity assumptions for this discussion and later applications of it (§15) is a non-trivial problem.

Consider a Markoff process x on $\mathbb{R}^{\ell}$ of the form

$$dx(t) = b(x(t),t)dt + dw(t),$$

where w is a Wiener process on $\mathbb{R}^{\ell}$ with diffusion coefficient ν (we write ν instead of D to avoid confusion with mean forward derivatives). Here b is a fixed smooth function on $\mathbb{R}^{\ell+1}$. The $w(t)-w(s)$ are independent of the $x(r)$ whenever $r \leq s$ and $r \leq t$, so that

$$Dx(t) = b(x(t),t).$$

A Markoff process with time reversed is again a Markoff process (see Doob [15, §6, p.83]), so we may define b_* by

$$D_*x(t) = b_*(x(t),t)$$

and w_* by

$$dx(t) = b_*(x(t),t)dt + dw_*(t).$$

Let f be a smooth function on $\mathbb{R}^{\ell+1}$. Then

$$f(x(t+\Delta t),t+\Delta t) - f(x(t),t) =$$

$$\frac{\partial f}{\partial t}(x(t),t)\Delta t + [x(t+\Delta t) - x(t)]\cdot\nabla f(x(t),t)$$

$$+ \frac{1}{2}\sum_{i,j} [x_i(t+\Delta t) - x_i(t)][x_j(t+\Delta t) - x_j(t)]\frac{\partial^2 f}{\partial x^i \partial x^j}(x(t),t)$$

$$+ o(\Delta t),$$

so that

(1) $$Df(x(t),t) = \left(\frac{\partial}{\partial t} + b\cdot\nabla + \nu\Delta\right)f(x(t),t).$$

Let ν_* be the diffusion coefficient of w_*. (A priori, ν_* might depend on x and t, but we shall see shortly that $\nu_*=\nu$.) Similarly, we find

(2) $$D_*f(x(t),t) = \left(\frac{\partial}{\partial t} + b_*\cdot\nabla - \nu_*\Delta\right)f(x(t),t).$$

If f and g have compact support in time, then Theorem 11.12 shows that

$$\int_{-\infty}^{\infty} EDf(x(t),t)\cdot g(x(t),t)dt = -\int_{-\infty}^{\infty} Ef(x(t),t)D_*g(x(t),t)dt;$$

that is,

$$\int_{-\infty}^{\infty}\int_{\mathbb{R}^\ell}\left(\frac{\partial}{\partial t} + b\cdot\nabla + \nu\Delta\right)f(x,t)\cdot g(x,t)\rho(x,t)dxdt =$$

$$\int_{-\infty}^{\infty}\int_{\mathbb{R}^\ell} f(x,t)\left(\frac{\partial}{\partial t} + b\cdot\nabla - \nu_*\Delta\right)g(x,t)\cdot\rho(x,t)dxdt.$$

For A a partial differential operator, let A^+ be its (Lagrange) adjoint with respect to Lebesgue measure on $\mathbb{R}^{\ell+1}$ and let A^* be its adjoint with respect to ρ times Lebesgue measure. Then $\int(Af)g\rho$ is equal to both $\int fA^+(g\rho)$ and $\int f(A^*g)\rho$, so that

$$A^* = \rho^{-1}A^+\rho.$$

Now

$$\left(\frac{\partial}{\partial t} + b\cdot\nabla + \nu\Delta\right)^{+} = -\frac{\partial}{\partial t} - b\cdot\nabla - \operatorname{div} b + \nu\Delta,$$

so that

$$\rho^{-1}\left(\frac{\partial}{\partial t} + b\cdot\nabla + \nu\Delta\right)^{+}\rho g = \rho^{-1}\left(-\frac{\partial}{\partial t} - b\cdot\nabla - \operatorname{div} b + \nu\Delta\right)(\rho g) =$$

$$-\frac{\partial g}{\partial t} - \rho^{-1}\frac{\partial \rho}{\partial t} g - b\cdot\nabla g - \rho^{-1} b\cdot(\operatorname{grad}\rho)g - (\operatorname{div} b)g$$

$$+ \rho^{-1}\nu((\Delta\rho)g + 2\operatorname{grad}\rho\cdot\operatorname{grad}\rho + \rho\Delta g).$$

Recall the Fokker-Planck equation

$$\frac{\partial \rho}{\partial t} = -\operatorname{div}(b\rho) + \nu\Delta\rho. \tag{3}$$

Using this we find

$$-\rho^{-1}\frac{\partial\rho}{\partial t} = \frac{\operatorname{div}(b\rho)}{\rho} - \nu\frac{\Delta\rho}{\rho} = \operatorname{div} b + b\cdot\frac{\operatorname{grad}\rho}{\rho} - \nu\frac{\Delta\rho}{\rho},$$

so we get

$$-\frac{\partial}{\partial t} - b_{*}\cdot\nabla + \nu_{*}\Delta = -\frac{\partial}{\partial t} - b\cdot\nabla + 2\nu\frac{\operatorname{grad}\rho}{\rho}\cdot\nabla + \nu\Delta.$$

Therefore, $\nu_{*}=\nu$ and $b_{*}=b-2\nu(\operatorname{grad}\rho)/\rho$. If we make the definition

$$u = \frac{b-b_{*}}{2},$$

we have

$$u = \nu\frac{\operatorname{grad}\rho}{\rho}.$$

We call u the <u>osmotic velocity</u> (cf. §4 Eq.(6)).

There is also a Fokker-Planck equation for time reversed:

$$\frac{\partial\rho}{\partial t} = -\operatorname{div}(b_{*}\rho) - \nu\Delta\rho. \tag{4}$$

If we define

$$v = \frac{b+b_{*}}{2},$$

we have the equation of continuity

$$\frac{\partial \rho}{\partial t} = -\operatorname{div}(v\rho),$$

obtained by averaging (3) and (4). We call v the <u>current velocity</u>.

Now

$$u = \nu \frac{\operatorname{grad} \rho}{\rho} = \nu \operatorname{grad} \log \rho.$$

Therefore,

$$\frac{\partial u}{\partial t} = \nu \operatorname{grad} \frac{\partial}{\partial t} \log \rho = \nu \operatorname{grad} \frac{\frac{\partial \rho}{\partial t}}{\rho} =$$

$$\nu \operatorname{grad} \frac{(-\operatorname{div}(v\rho))}{\rho} = -\nu \operatorname{grad}(\operatorname{div} v + v \cdot \frac{\operatorname{grad} \rho}{\rho}) =$$

$$-\nu \operatorname{grad} \operatorname{div} v - \operatorname{grad} v \cdot u.$$

That is,

(5) $$\frac{\partial u}{\partial t} = -\nu \operatorname{grad} \operatorname{div} v - \operatorname{grad} v \cdot u.$$

Finally, from (1) and (2),

$$Db_*(x(t),t) = \frac{\partial}{\partial t} b_*(x(t),t) + b \cdot \nabla b_*(x(t),t) + \nu \Delta b_*(x(t),t),$$

$$D_* b(x(t),t) = \frac{\partial}{\partial t} b(x(t),t) + b_* \cdot \nabla b(x(t),t) - \nu \Delta b(x(t),t),$$

so that the mean acceleration (as defined in §11, Eq.(15)) is given by $a(x(t),t)$ where

$$a = \frac{\partial}{\partial t}\left(\frac{b+b_*}{2}\right) + \frac{1}{2} b \cdot \nabla b_* + \frac{1}{2} b_* \cdot \nabla b + \nu \Delta\left(\frac{b-b_*}{2}\right).$$

That is,

(6) $$\frac{\partial v}{\partial t} = a + u \cdot \nabla u - v \cdot \nabla v + \nu \Delta u.$$

§14. Remarks on quantum mechanics

In discussing physical theories of Brownian motion we have seen that physics has interesting ideas and problems to contribute to probability theory. Probabilities also play a fundamental rôle in quantum mechanics, but the notion of probability enters in a new way which is foreign both to classical mechanics and to mathematical probability theory. A mathematician interested in probability theory should become familiar with the peculiar concept of probability in quantum mechanics.

We shall discuss quantum mechanics from the point of view of the rôle of probabilistic concepts in it, limiting ourselves to the non-relativistic quantum mechanics of systems of finitely many degrees of freedom. This theory was discovered in 1925-1926. Its principal features were established quickly, and it has changed very little in the last forty years.

Quantum mechanics originated in an attempt to solve two puzzles: the discrete atomic spectra and the dual wave-particle nature of matter and radiation. Spectroscopic data were interpreted as being evidence for the fact that atoms are mechanical systems which can exist in stationary states only for a certain discrete set of energies.

There have been many discussions of the two-slit thought experiment illustrating the dual nature of matter; e.g., [28, §12] and [29, Ch.1]. Here we merely recall the bare facts: A particle issues from x in the figure, passes through the doubly-slitted screen in the middle, and hits the screen on the right, where its

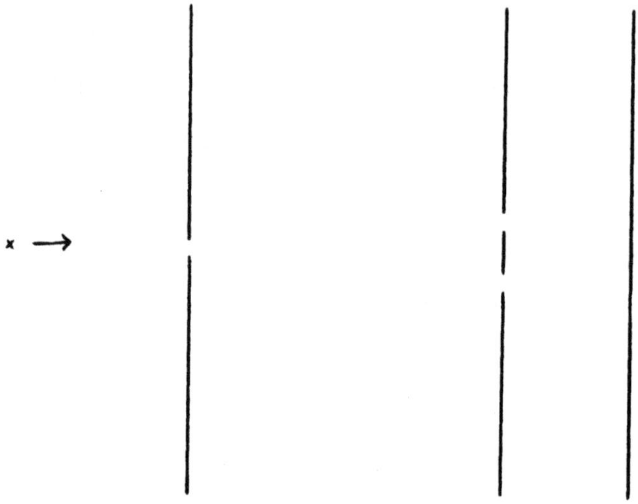

Fig. 3

position is recorded. Particle arrivals are sharply localized indivisible events, but despite this the probability of arrival shows a complicated diffraction pattern typical of wave motion. If one of the holes is closed, there is no interference pattern. If an observation is made (using strong light of short wave length) to see which of the two slits the particle went through, there is again no interference pattern.

The founders of quantum mechanics may be divided into two groups: the reactionaries (Planck, Einstein, de Broglie, Schrödinger) and the radicals (Bohr, Heisenberg, Born, Jordan, Dirac). Correspondingly, quantum mechanics was discovered in two apparently different forms: wave mechanics and matrix mechanics. (Heisenberg's original term was "quantum mechanics," and "matrix mechanics" is used when one wishes to distinguish it from Schrödinger's wave mechanics.)

In 1900 Planck introduced the quantum of action h and in 1905 Einstein postulated particles of light with energy $E=h\nu$ (ν the frequency). We give no details as we shall not discuss radiation. In 1924, while a graduate student at the Sorbonne, Louis de Broglie put the two formulas $E=mc^2$ and $E=h\nu$ together and invented matter waves. The wave nature of matter received experimental confirmation in the Davisson-Germer electron diffraction experiment of 1927, and theoretical support by the work of Schrödinger in 1926. De Broglie's thesis committee included Perrin, Langevin, and Elie Cartan. Perhaps Einstein heard of de Broglie's work from Langevin. In any case, Einstein told Born, "Read it; even though it looks crazy it's solid," and he published comments on de Broglie's work which Schrödinger read.

Suppose, with Schrödinger, that we have a particle (say an electron) of mass m in a potential V. Here V is a real function on $\mathbb{R}^3$ representing the potential energy. Schrödinger attempted to describe the motion of the electron by means of a quantity ψ subject to a wave equation. He was led to the hypothesis that a stationary state vibrates according to the equation

$$\frac{\hbar^2}{2m}\Delta\psi + (E-V)\psi = 0, \tag{1}$$

where $\hbar$ is Planck's constant h over 2π and E (with the dimensions of energy) plays the rôle of an eigenvalue.

This equation is similar to the wave equation for a

vibrating eleastic fluid contained in a given enclosure, except that V is not a constant. Schrödinger was struck by another difference [30, p.12]:

"A simplification in the problem of the 'mechanical' waves (as compared with the fluid problem) consists in the absence of boundary conditions. I thought the latter simplification fatal when I first attacked these questions. Being insufficiently versed in mathematics, I could not imagine how proper vibration frequencies could appear <u>without</u> boundary conditions."

Despite these misgivings, Schrödinger found the eigenvalues and eigenfunctions of (1) for the case of the hydrogen atom, $V=-e^2/r$ where e is the charge of the electron (and -e is the charge of the nucleus) and $r^2=x^2+y^2+z^2$. The eigenvalues corresponded precisely to the known discrete energy levels of the hydrogen atom.

This initial triumph, in which discrete energy levels appeared for the first time in a natural way, was quickly followed by many others. Before the year of 1926 was out, Schrödinger reprinted six papers on wave mechanics in book form [30]. A young lady friend remarked to him (see the preface to [30]): "When you began this work you had no idea that anything so clever would come out of it, had you?" With this remark Schrödinger "wholeheartedly agreed (with due qualification of the flattering adjective)."

Shortly before Schrödinger made his discovery, the matrix mechanics of Heisenberg appeared. In this theory one constructs six infinite matrices q_k and p_j (j,k=1,2,3) satisfying the commutation relations

$$p_j q_k - q_k p_j = \frac{\hbar}{i}\delta_{jk}$$

and diagonalizes the matrix $H=p^2/2m+V(q)$. Schrödinger remarks [30, p.46]:

"My theory was inspired by L. de Broglie, Ann. de Physique (10)3, p.22, 1925 (Theses, Paris, 1924), and by brief, yet infinitely far-seeing remarks of A. Einstein, Berl. Ber., 1925, p.9 et seq. I did not at all suspect any relation to Heisenberg's theory at the beginning. I naturally knew about his theory, but was discouraged, if not repelled, by what appeared to me as very difficult methods of transcendental algebra, and by the want of perspecuity (Anschaulichkeit)."

The remarkable thing was that where the two theories disagreed with the old quantum theory of Bohr, they agreed with each other (and with experiment!). Schrödinger quickly discovered the mathematical equivalence of the two theories, based on letting q_k correspond to the operator of multiplication by the coordinate function x_k and letting p_j correspond to the operator $(\hbar/i)\partial/\partial x_j$ (see the fourth paper in [30]).

Schrödinger maintained (and most physicists agree) that the mathematical equivalence of two physical theories is not the same as their physical equivalence, and went on to describe a possible physical interpretation of the wave function ψ. According to this interpretation an electron with wave function ψ is not a localized particle but a smeared out distribution of electricity with charge density $e\rho$ and electric current ej, where

$$\rho = |\psi|^2, \qquad j = \frac{i\hbar}{2m}(\psi \operatorname{grad} \bar{\psi} - \bar{\psi} \operatorname{grad} \psi).$$

(The quantities ρ and j determine ψ except for a multiplicative factor of absolute value one.) This interpretation works very well for a single electron bound in an atom, provided one neglects the self-repulsion of the smeared out electron. However, when there are n electrons, ψ is a function on configuration space $\mathbb{R}^{3n}$ rather than coordinate space $\mathbb{R}^3$, which makes the interpretation of ψ as a physically real object very difficult. Also, for free electrons ψ, and consequently ρ, spreads out more and more as time goes on; yet the arrival of electrons at a scintillation screen is always signalled by a sharply localized flash, never by a weak, spread out flash. These objections were made to Schrödinger's theory when he lectured on it in Copenhagen, and he reputedly said he wished he had never invented the theory.

The accepted interpretation of the wave function ψ was put forward by Born [31], and quantum mechanics was given its present form by Dirac [32] and von Neumann [33]. Let us briefly describe quantum mechanics, neglecting superselection rules.

To each physical system there corresponds a Hilbert space $\mathcal{H}$. To every state (also called pure state) of the system there corresponds an equivalence class of unit vectors in $\mathcal{H}$, where ψ_1 and ψ_2 are called equivalent if $\psi_1 = a\psi_2$ for some complex number a of absolute value one. (Such an equivalence class, which is a circle, is frequently called a ray.) The correspondence between states and rays is one-to-one. To each observable of the system there corresponds a self-adjoint operator, and the correspondence is again one-to-one. The development of the system in time is described by a family of unitary operators $U(t)$ on $\mathcal{H}$. There are two ways of thinking about this. In the Schrödinger picture, the state of the system changes with time — $\psi(t)=U(t)\psi_0$, where ψ_0 is the state at time 0, and observables do not change with time. In the Heisenberg picture, observables change with time — $A(t)=U(t)^{-1}A_0U(t)$, and the state does not change with time. The two pictures are equivalent, and it is a matter of convention which is used. For an isolated physical system, the dynamics is given by $U(t)=\exp(-(i/\hbar)Ht)$, where H, the Hamiltonian, is the self-adjoint operator representing the energy of the system.

It may happen that one does not know the state of the physical system, but merely that it is in state ψ_1 with probability w_1, state ψ_2 with probability w_2, etc., where $w_1+w_2+\ldots=1$. This is called a mixture (impure state), and we shall not describe its mathematical representation further.

The important new notion is that of a superposition of states. Suppose that we have two states ψ_1 and ψ_2. The number $|(\psi_1,\psi_2)|^2$ does not depend on the choice of representatives of the rays and lies between 0 and 1. Therefore, it may be regarded as a probability. If we know that the system is in the state ψ_1 and we perform an experiment to see whether or not the system is in the state ψ_2, then $|(\psi_1,\psi_2)|^2$ is the probability of finding that the system is indeed in the state ψ_2. We may write

$$\psi_1 = (\psi_2,\psi_1)\psi_2 + (\psi_3,\psi_1)\psi_3$$

where ψ_3 is orthogonal to ψ_2. We say that ψ_1 is a superposition of the states ψ_2 and ψ_3. Consider the mixture which is in the state ψ_2 with probability $|(\psi_2,\psi_1)|^2$ and in the state ψ_3 with probability $|(\psi_3,\psi_1)|^2$. Then ψ_1 and the mixture have equal probabilities of being found in the states ψ_2 and ψ_3, but they are quite different. For example, ψ_1 has the probability $|(\psi_1,\psi_1)|^2=1$ of being found in the state ψ_1, whereas the mixture has only the probability $|(\psi_2,\psi_1)|^4+|(\psi_3,\psi_1)|^4$ of being found in the state ψ_1.

A superposition represents a number of different possibilities, but unlike a mixture the different possibilities may interfere. Thus in the two-slit experiment, the particle is in a superposition of states of passing through the top slit and the bottom slit, and the interference of these possibilities leads to the diffraction pattern. If we look to see which slit the particle comes through then the particle will be in a mixture of states of passing through the top slit and the bottom slit and there will be no diffraction pattern.

If the system is in the state ψ and A is an observable with spectral projections E_λ then $(\psi,E_\lambda\psi)$ is the probability that if we perform an experiment to determine the value of A we will obtain a result $\leq\lambda$. Thus $(\psi,A\psi)=\int\lambda(\psi,dE_\lambda\psi)$ is the expected value of A in the state ψ. (The left hand side is meaningful if ψ is in the domain of A; the integral on the right hand side converges if ψ is merely in the domain of $|A|^{\frac{1}{2}}$.) The observable A has the value λ with certainty if and only if ψ is an eigenvector of A with eigenvalue λ, $A\psi=\lambda\psi$.

Thus quantum mechanics differs from classical mechanics in not requiring every observable to have a sharp value in every (pure) state. Furthermore, it is in general impossible to find a state such that two given observables have sharp values. Consider the position operator q and the momentum operator p for a particle with one degree of freedom, and let ψ be in the domain of p^2, q^2, pq, and qp. Then $(\psi,p^2\psi)-(\psi,p\psi)^2=(\psi,(p-(\psi,p\psi))^2\psi)$ is the variance of the observable p in the state ψ and its square root is the standard deviation, which physicists frequently call the dispersion and denote by Δp. Similarly for $(\psi,q^2\psi)-(\psi,q\psi)^2$. If we use the commutation relation

$$(2) \qquad (pq - qp)\psi = \frac{\hbar}{i}\psi$$

we find that for all real α,

$$\begin{aligned} 0 \le ((\alpha q + ip)\psi, (\alpha q + ip)\psi) = \\ \alpha^2(\psi,q^2\psi) - i\alpha(\psi,(pq - qp)\psi) + (\psi,p^2\psi) = \\ \alpha^2(\psi,q^2\psi) - \alpha\hbar + (\psi,p^2\psi). \end{aligned}$$

Since this is positive for all real α, the discriminant must be negative,

$$(3) \qquad \hbar^2 - 4(\psi,q^2\psi)(\psi,p^2\psi) \le 0.$$

The commutation relation (2) continues to hold if we replace p by $p-(\psi,p\psi)$ and q by $q-(\psi,q\psi)$, so (3) continues to hold after this replacement. That is,

$$(4) \qquad \Delta q \Delta p \ge \frac{\hbar}{2}.$$

This is the well-known proof of the Heisenberg uncertainty relation. The great importance of Heisenberg's discovery, however, was not the formal deduction of this relation but the presentation of arguments which showed, in an endless string of cases, that the relation (4) must hold on physical grounds independently of the formalism.

Thus probabilistic notions are central in quantum mechanics. Given the state ψ, the observable A may be regarded as a random variable on the probability space consisting of the real line with the measure $(\psi, dE_\lambda \psi)$, where the E_λ are the spectral projections of A. Similarly, any number of commuting self-adjoint operators may be regarded as random variables on a probability space. (Two self-adjoint operators are said to commute if their spectral projections commute.) But, and it is this which makes quantum mechanics so radically different from classical theories, the set of all observables of the system in a given state cannot be regarded as a set of random variables on a probability space. For example, the formalism of quantum mechanics does not allow the possibility of p and q both having sharp values even if the putative sharp values are unknown.

For a while it was thought by some that there might be "hidden variables" - that is, a more refined description of the state of a system - which would allow all observables to have sharp values if a complete description of the system were known. Von Neumann [33] showed, however, that any such theory would be a departure from quantum mechanics rather than an extension of it. It follows from von Neumann's theorem that the set of all self-adjoint operators in a given state cannot be regarded as a family of random variables on a probability space. Here is another result along these lines.

Theorem 14.1. *Let $A=(A_1,\ldots,A_n)$ be an n-tuple of operators on a Hilbert space $\mathcal{H}$ such that for all x in $\mathbb{R}^n$,*

$$x\cdot A = x_1A_1 + \ldots + x_nA_n$$

is essentially self-adjoint. Then either the $A_1,\ldots,A_n$ commute or there is a ψ in $\mathcal{H}$ with $\|\psi\|=1$ such that there do not exist random variables $\alpha_1,\ldots,\alpha_n$ on a probability space with the property that for all x in $\mathbb{R}^n$ and λ in $\mathbb{R}$,

$$\Pr\{x\cdot\alpha \geq \lambda\} = (\psi,E_\lambda(x\cdot A)\psi),$$

where $x\cdot\alpha=x_1\alpha_1+\ldots+x_n\alpha_n$ and the $E_\lambda(x\cdot A)$ are the spectral projections of the closure of $x\cdot A$.

In other words, n observables may be regarded as random variables, in all states, if and only if they commute.

Proof. We shall not distinguish notationally between $x\cdot A$ and its closure.

Suppose that for each unit vector ψ in $\mathcal{H}$ there is such an n-tuple α of random variables, and let μ_ψ be the probability distribution of α on $\mathbb{R}^n$. That is, for each Borel set B in $\mathbb{R}^n$, $\mu_\psi(B)=\Pr\{\alpha\in B\}$. If we integrate first over the hyperplanes orthogonal to x, we find that

$$\int_{\mathbb{R}^n} e^{ix\cdot\xi}d\mu_\psi(\xi) = \int_{-\infty}^{\infty} e^{i\lambda}d\Pr\{x\cdot\alpha\geq\lambda\}$$

$$= \int_{-\infty}^{\infty} e^{i\lambda}(\psi,dE_\lambda(x\cdot A)\psi) = (\psi,e^{ix\cdot A}\psi).$$

Thus the measure μ_ψ is the Fourier transform of $(\psi,e^{ix\cdot A}\psi)$. By the polarization identity, if φ and ψ are in $\mathcal{H}$ there is a complex measure $\mu_{\varphi\psi}$ such that $\mu_{\varphi\psi}$ is the Fourier transform of $(\varphi, e^{ix\cdot A}\psi)$ and $\mu_{\psi\psi}=\mu_\psi$. For any Borel set B in $\mathbb{R}^n$ there is a unique operator $\mu(B)$ such that $(\varphi,\mu(B)\psi)=\mu_{\varphi\psi}(B)$, since $\mu_{\varphi\psi}$ depends linearly on ψ and antilinearly on φ. Thus we have

$$\int_{\mathbb{R}^n} e^{ix\cdot\xi}(\varphi, d\mu(\xi)\psi) = (\varphi, e^{ix\cdot A}\psi).$$

The operator $\mu(B)$ is positive since μ_ψ is a positive measure. Consequently, if we have a finite set of elements ψ_j of $\mathcal{H}$ and corresponding points x_j of $\mathbb{R}^n$, then

$$\sum_{j,k} (\psi_k, e^{i(x_j-x_k)\cdot A}\psi_j) =$$

$$\sum_{j,k} \int_{\mathbb{R}^n} e^{i(x_j-x_k)\cdot\xi}(\psi_k, d\mu(\xi)\psi_j) =$$

$$\int_{\mathbb{R}^n} (\psi(\xi), d\mu(\xi)\psi(\xi)) \geq 0,$$

where

$$\psi(\xi) = \sum_j e^{ix_j\cdot\xi}\psi_j.$$

Furthermore, $e^{i0\cdot A}=1$ and $e^{i(-x)\cdot A}=(e^{ix\cdot A})^*$. Under these conditions, the theorem on unitary dilations of Nagy [34, Appendix, p.21] implies that there is a Hilbert space $\mathcal{K}$ containing $\mathcal{H}$ and a unitary representation $x\to U(x)$ of $\mathbb{R}^n$ on $\mathcal{K}$ such that, if E is the orthogonal projection of $\mathcal{K}$ onto $\mathcal{H}$, then

$$EU(x)\psi = e^{ix\cdot A}\psi$$

for all x in $\mathbb{R}^n$ and all ψ in $\mathcal{H}$. Since $e^{ix\cdot A}$ is already unitary,

$$\|U(x)\psi\| = \|e^{ix\cdot A}\psi\| = \|\psi\|,$$

so that $\|EU(x)\psi\|=\|U(x)\psi\|$. Consequently, $EU(x)\psi=U(x)\psi$ and each $U(x)$ maps $\mathcal{H}$ into itself, so that $U(x)\psi = e^{ix\cdot A}\psi$ for all ψ in $\mathcal{H}$. Since $x \to U(x)$ is a unitary representation of the commutative group $\mathbb{R}^n$, the $e^{ix\cdot A}$ all commute, and consequently the A_j commute. QED.

Quantum mechanics forced a major change in the notion of reality. The position and momentum of a particle could no longer be thought of as properties of the particle. They had no real existence before measurement, and the measurement of the one with a given accuracy precluded the measurement of the other with too great an accuracy, in accordance with the uncertainty principle. This point of view was elaborated by Bohr under the slogan of "complementarity," and Heisenberg wrote a book [35] explaining the physical basis of the new theory.

At the Solvay Congress in 1927, Einstein was very quiet, but when pressed objected that ψ could not be the complete description of the state. For example, the wave function in Fig. 4 would have axial symmetry, but the place of arrival of an individual particle on the hemispherical screen does not have this symmetry. The answer of quantum mechanics is that the symmetrical wave function ψ describes the state of the system before a measurement is made, but the act of measurement changes ψ.

To understand the rôle of probability in quantum mechanics it is necessary to discuss measurement. The quantum theory of measurement was created by von Neumann [33]. A very readable summary is the book by London and Bauer [36]. See also two papers of Wigner [37], [38], which we follow now.

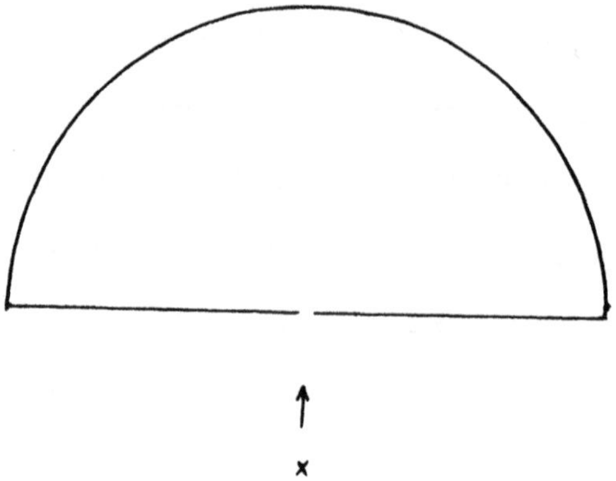

Fig. 4

Consider a physical system with wave function ψ in a state of superposition of the two orthogonal states ψ_1 and ψ_2, so that $\psi=\alpha_1\psi_1+\alpha_2\psi_2$. We want to perform an experiment to determine whether it is in the state ψ_1 or the state ψ_2. (We know that the probabilities are respectively $|\alpha_1|^2$ and $|\alpha_2|^2$, but we want to know which it is in.)

If A is any observable, the expected value of A is

$$(\alpha_1\psi_1+\alpha_2\psi_2, A(\alpha_1\psi_1+\alpha_2\psi_2)) =$$
$$|\alpha_1|^2(\psi_1,A\psi_1)+|\alpha_2|^2(\psi_2,A\psi_2)+\bar{\alpha}_1\alpha_2(\psi_1,A\psi_2)+\bar{\alpha}_2\alpha_1(\psi_2,A\psi_1).$$

Suppose now we couple the system to an apparatus designed to measure whether the system is in the state ψ_1 or ψ_2, and that after the interaction the system plus apparatus is in the state

$$\varphi = \alpha_1 \psi_1 \otimes \chi_1 + \alpha_2 \psi_2 \otimes \chi_2,$$

where χ_1 and χ_2 are orthogonal states of the apparatus. If A is any observable pertaining to the system alone, then

$$(\varphi, A \otimes 1\varphi) = |\alpha_1|^2(\psi_1, A\psi_1) + |\alpha_2|^2(\psi_2, A\psi_2).$$

Thus, after the interaction with the apparatus, the system behaves like a mixture of ψ_1 and ψ_2 rather than a superposition. It is in this sense that the interaction with the apparatus is a measurement of whether the system is in the state ψ_1 or ψ_2.

However, observe that knowing the state of the system plus apparatus after the interaction tells us nothing about which state the system is in! The state φ is a complete description of the system plus apparatus. It is causally determined by ψ and the initial state of the apparatus, according to the Schrödinger equation governing the interaction. Thus letting the system interact with an apparatus can never give us more information.

If we knew that after the interaction the apparatus is in the state χ_1 we would know that the system is in the state ψ_1. But how do we tell whether the apparatus is in state χ_1 or χ_2? We might couple it to another apparatus, but this threatens an infinite regress.

In practice, however, the apparatus is macroscopic, like a spot on a photographic plate or the pointer on a meter, and I merely look to see which state it is in, χ_1 or χ_2. After I have become aware of the state χ_1 or χ_2, the act of measurement is complete.

If I see that the apparatus is in state χ_1, the system plus apparatus is in state $\psi_1 \otimes \chi_1$ and the system is in state ψ_1. (As we already knew, this will happen with probability $|\alpha_1|^2$.) Similarly for χ_2. After the interaction but before awareness has dawned in me, the state of the system plus apparatus is $\alpha_1\psi_1 \otimes \chi_1 + \alpha_2\psi_2 \otimes \chi_2$; the instant I become aware of the state of the apparatus, the state of the system plus apparatus is either $\psi_1 \otimes \chi_1$ or $\psi_2 \otimes \chi_2$. Thus the state can change in two ways: continuously, linearly, and causally by means of the Schrödinger equation or abruptly, nonlinearly, and probabilistically by means of my consciousness. The latter change is called "reduction of the wave packet." Concerning the reduction of the wave packet, Wigner [39] writes:

"This takes place whenever the result of an observation enters the consciousness of the observer - or, to be even more painfully precise, my own consciousness, since I am the only observer, all other people being only subjects of my observations."

This theory is known as the orthodox theory of measurement in quantum mechanics. The word "orthodox" is well chosen: one suspects that many practicing physicists are not entirely orthodox in their beliefs.

Of those founders of quantum mechanics whom we labelled "reactionaries," none permanently accepted the orthodox interpretation of quantum mechanics. Schrödinger writes [40, p.16]:

"For it must have given to de Broglie the same shock and disappointment as it gave to me, when we learnt that a sort of transcendental, almost psychial interpretation of the wave phenomenon had been put forward, which was very soon hailed by the majority of leading theorists as the only one reconcilable with experiments, and which has now become the orthodox creed, accepted by almost everybody, with a few notable exceptions."

The literature on the interpretation of quantum mechanics contains much of interest, but I shall discuss only three memorable paradoxes: the paradox of Schrödinger's cat [41, p.812], the paradox of the nervous student [42] [43] [44], and the paradox of Wigner's friend [38]. The original accounts of these paradoxes make very lively reading.

One is inclined to accept rather abstruse descriptions of electrons and atoms, which one has never seen. Consider, however, a cat which is enclosed in a vault with the following infernal machine, located out of reach of the cat. A small amount of a radioactive substance is present, with a half-life such that the probability of a single decay in an hour is about one half. If a radioactive decay occurs, a counter activates a device which breaks a phial of prussic acid, killing the cat. The only point of this paradox is to consider what, according to quantum mechanics, is a <u>complete description</u> of the state of affairs in the vault at the end of an hour. One cannot say that the cat is alive or dead, but that the state of it and the infernal machine is described by a superposition of various states containing dead and alive cats, in which the cat variables are mixed up with the machine variables. This precise state of affairs is the ineluctable outcome of the initial conditions. Unlike most thought experiments, this one could actually be performed, were it not inhumane.

The first explanations of the uncertainty principle (see [35]) made it seem the natural result of the fact that, for example, observing the position of a particle involves giving the particle a kick and thereby changing its momentum. Einstein, Podolsky, and Rosen [42] showed that the situation is not that simple. Consider two particles with one degree of freedom. Let x_1 and p_1 denote the position and momentum operators of the first, x_2 and p_2 those of the second. Now $x=x_1-x_2$ and $p=p_1+p_2$ are commuting operators and so may simultaneously have sharp values, say x' and p' respectively. Suppose that x' is very large, so that the two particles are very far apart. Then we may measure x_2, obtaining the value x_2', say, without in any way affecting the first particle. A measurement of x_1 then must give $x'+x_2'$. Since the measurement of x_2 cannot have affected the first particle (which is very far away), there must have been something about the condition of the first particle which meant that x_1 if measured would give the value $x_1'=x'+x_2'$. Similarly for position measurements. To quote Schrödinger [44]:

"Yet since I can predict <u>either</u> x_1' <u>or</u> p_1' without interfering with system No.1 and since system No.1, like a scholar in an examination, cannot possibly know which of the two questions I am going to ask it first: it so seems that our scholar is prepared to give the right answer to the <u>first</u> question he is asked, <u>anyhow.</u> Therefore he must know both answers; which is an amazing knowledge, quite irrespective of the fact that after having given his first answer our scholar is invariably so disconcerted or tired out, that all the following answers are 'wrong'."

The paradox of Wigner's friend must be told in the first person. There is a physical system in the state $\psi=\alpha_1\psi_1+\alpha_2\psi_2$ which, if in the state ψ_1, produces a flash, and if in the state ψ_2, does not. In the description of the measurement process given earlier, for the apparatus I substitute my friend. After the interaction of system and friend they are in the state $\varphi=\alpha_1\psi_1\otimes\chi_1+\alpha_2\psi_2\otimes\chi_2$. I ask my friend if he saw a flash. If I receive the answer "yes" the state changes abruptly (reduction of the wave packet) to $\psi_1\otimes\chi_1$; if I receive the answer "no" the state changes to $\psi_2\otimes\chi_2$. Now suppose I ask my friend, "What did you feel about the flash before I asked you?" He will answer that he already knew that there was (or was not) a flash before I asked him. If I accept this, I must accept that the state was $\psi_1\otimes\chi_1$ (or $\psi_2\otimes\chi_2$), rather than φ, in violation of the laws of quantum mechanics. One possibility is to deny the existence of consciousness in my friend (solipsism). Wigner prefers to believe that the laws of quantum mechanics are violated by his friend's consciousness.

References

[29]. R. P. Feynman and A. R. Hibbs, Quantum Mechanics and Path Integrals, McGraw-Hill, New York, 1965.

[30]. Erwin Schrödinger, Collected Papers on Wave Mechanics, translated by J. F. Shearer and W. M. Deans, Blackie & son limited, London, 1928.

[31]. Max Born, Zur Quantenmechanik der Stossvorgänge, Zeitschrift für Physik 37(1926), 863-867, 38, 803-827.

[32]. P. A. M. Dirac, The Principles of Quantum Mechanics, Oxford, 1930.

[33]. John von Neumann, Mathematical Foundations of Quantum Mechanics, translated by R. T. Beyer, Princeton University Press, Princeton, 1955.

[34]. Frigyes Riesz and Béla Sz.-Nagy, Functional Analysis, 2nd Edition, translated by L. F. Boron, with Appendix: Extensions of Linear Transformations in Hilbert Space which Extend Beyond This Space, by B. Sz.-Nagy, Frederick Ungar Publishing Co., New York, 1960.

[35]. Werner Heisenberg, The physical principles of the quantum theory, translated by Carl Eckhart and Frank C. Hoyt, Dover Publications, New York, 1930.

[36]. F. London and E. Bauer, La théorie de l'observation en mécanique quantique, Hermann et Cie., Paris, 1939.

[37]. Eugene P. Wigner, The problem of measurement, American Journal of Physics 31(1963), 6-15.

[38]. Eugene P. Wigner, Remarks on the mind-body question, pp.284-302 in The Scientist Speculates, edited by I. J. Good, 1962.

[39]. Eugene P. Wigner, Two kinds of reality, The Monist 48(1964), 248-264.

[40]. Erwin Schrödinger, The meaning of wave mechanics, pp.16-30 in Louis de Broglie physicien et penseur, edited by André George, éditions Albin Michel, Paris, 1953.

[41]. Erwin Schrödinger, Die gegenwärtige Situation in der Quantenmechanik, published as a serial in Naturwissenschaften 23(1935), 807-812, 823-828, 844-849.

[42]. A. Einstein, B. Podolsky and N. Rosen, Can quantum-mechanical description of reality be considered complete?, Physical Review 47(1935), 777-780.

[43]. N. Bohr, Can quantum-mechanical description of reality be considered complete?, Physical Review 48(1935), 696-702.

[44]. E. Schrödinger, Discussion of probability relations between separated systems, Proceedings of the Cambridge Philosophical Society 31(1935), 555-563.

An argument against hidden variables which is much more incisive than von Neumann's is presented in a forthcoming paper:

[45]. Simon Kochen and E. P. Specker, The problem of hidden variables in quantum mechanics, Journal of Mathematics and Mechanics, to appear.

§15. Brownian motion in the aether

Let us try to see whether some of the puzzling physical phenomena which occur on the atomic scale can be explained by postulating a kind of Brownian motion which agitates all particles of matter, even particles far removed from all other matter. It is not necessary to think of a material model of the aether and to imagine the cause of the motion to be bombardment by grains of the aether. Let us, for the present, leave the cause of the motion unanalyzed and return to Robert Brown's conception of matter as composed of small particles which exhibit a rapid irregular motion having its origin in the particles themselves (rather like Mexican jumping beans).

We cannot suppose that the particles experience friction in moving through the aether as this would imply that uniform rectilinear motion could be distinguished from absolute rest. Consequently, we cannot base our discussion on the Langevin equation.

We shall assume that every particle performs a Markoff process of the form

$$dx(t) = b(x(t),t)dt + dw(t), \tag{1}$$

where w is a Wiener process on $\mathbb{R}^3$, with $w(t)-w(s)$ independent of $x(r)$ whenever $r \leq s \leq t$. Macroscopic bodies do not appear to move like this, so we shall postulate that the diffusion coefficient ν is inversely proportional to the mass m of the particle. We write it as

$$\nu = \frac{\hbar}{2m}.$$

The constant $\hbar$ has the dimensions of action. If $\hbar$ is of the order of Planck's constant h then the effect of the Wiener process would indeed not be noticeable for macroscopic bodies but would be relevant on the atomic scale. (Later we will see that $\hbar=h/2\pi$.) The kinematical assumption (1) is non-relativistic, and the theory we are proposing is meant only as an approximation valid when relativistic effects may safely be neglected.

We have already (§13) studied the kinematics of such a process. We let b_* be the mean backward velocity, $u=(b-b_*)/2$, $v=(b+b_*)/2$. By (5) and (6) of §13,

$$(2)\qquad \begin{aligned} \frac{\partial u}{\partial t} &= -\frac{\hbar}{2m}\,\text{grad div } v - \text{grad } v\cdot u, \\ \frac{\partial v}{\partial t} &= a - v\cdot\nabla v + u\cdot\nabla u + \frac{\hbar}{2m}\,\Delta u, \end{aligned}$$

where a is the mean acceleration.

Suppose that the particle is subject to an external force F. We make the dynamical assumption F=ma, and substitute F/m for a in (2). (This agrees with what is done in the Ornstein-Uhlenbeck theory of Brownian motion with friction (§12).)

Consider the case when the external force is derived from a potential V, which may be time-dependent, so that $F(x,t)=-\text{grad } V(x,t)$. Then (2) becomes

$$(3)\qquad \begin{aligned} \frac{\partial u}{\partial t} &= -\frac{\hbar}{2m}\,\text{grad div } v - \text{grad } v\cdot u \\ \frac{\partial v}{\partial t} &= \frac{1}{m}\,\text{grad } V - v\cdot\nabla v + u\cdot\nabla u + \frac{\hbar}{2m}\,\Delta u. \end{aligned}$$

If $u_0(x)$ and $v_0(x)$ are given, we have an initial value problem: to solve the system (3) of coupled non-linear partial differential equations subject to $u(x,0)=u_0(x)$, $v(x,0)=v_0(x)$ for all x in $\mathbb{R}^3$. Notice that when we do this we are not solving the equations of motion of the particle. We are merely finding what stochastic process the particle obeys, with the given force and the given initial osmotic and current velocities. Once u and v are known, b, b_*, and ρ are known, and so the Markoff process is known.

It would be interesting to know the general solution of the initial value problem (3). However, I can only solve it with the additional assumption that v is a gradient. (We already know (§13) that u is a gradient.) A solution of the problem without this assumption would seem to correspond to finding the Markoff process of the particle when the containing fluid, the aether, is in motion.

Let $R=\frac{1}{2}\log \rho$. Then we know (§13) that

$$\text{grad } R = \frac{m}{\hbar}u. \tag{4}$$

Assuming that v is also a gradient, let S be such that

$$\text{grad } S = \frac{m}{\hbar}v. \tag{5}$$

Then S is determined, for each t, up to an additive constant.

It is remarkable that the change of dependent variable

$$\psi = e^{R+iS} \tag{6}$$

transforms (3) into a linear partial differential equation; in fact, into the Schrödinger equation

$$\frac{\partial\psi}{\partial t} = i\frac{\hbar}{2m}\Delta\psi - i\frac{1}{\hbar}V\psi + i\alpha(t)\psi. \tag{7}$$

(Since the integral of $\rho=\bar{\psi}\psi$ is independent of t, if (7) holds at all then $\alpha(t)$ must be real. By choosing, for each t, the arbitrary constant in S appropriately we may arrange for $\alpha(t)$ to be 0.)

To prove (7), we compute the derivatives and divide by ψ, finding

$$\frac{\partial R}{\partial t} + i\frac{\partial S}{\partial t} = i\frac{\hbar}{2m}(\Delta R + i\Delta S + [\text{grad}(R + iS)]^2) - i\frac{1}{\hbar}V + i\alpha(t).$$

Taking gradients and separating real and imaginary parts, we see that this is equivalent to the pair of equations

$$\frac{\partial u}{\partial t} = -\frac{\hbar}{2m}\Delta v - \text{grad}\ v\cdot u,$$

$$\frac{\partial v}{\partial t} = \frac{\hbar}{2m}\Delta u + \frac{1}{2}\text{grad}\ u^2 - \frac{1}{2}\text{grad}\ v^2 - \frac{1}{m}\text{grad}\ V.$$

Since u and v are gradients, this is the same as (3).

Conversely, if ψ satisfies the Schrödinger equation (7) and we define R, S, u, v by (6), (4), and (5), then u and v satisfy (3). Note that u becomes singular when $\psi = 0$.

Is it an accident that Markoff processes of the form (1) with the dynamical law F=ma are formally equivalent to the Schrödinger equation? As a test, let us consider the motion of a particle in an external electromagnetic field. Let, as is customary, A be the vector potential, φ the scalar potential, E the electric field strength, H the magnetic field strength, and c the speed of light. Then

(8) $$H = \text{curl}\ A,$$

(9) $$E + \frac{1}{c}\frac{\partial A}{\partial t} = -\text{grad}\ \varphi.$$

The Lorentz force on a particle of charge e is

(10) $$F = e(E + \frac{1}{c}v\times H),$$

where v is the classical velocity. We adopt (10) as the force on a particle undergoing the Markoff process (1) with v the current velocity. We do this because the force should be invariant under time inversion $t \to -t$, and indeed $H \to -H$, $v \to -v$ (while $u \to u$) under time inversion. As before, we substitute F/m for a in (2). Now, however, we assume the generalized momentum $mv + eA/c$ to be a gradient. (This is a gauge invariant assumption.) Letting grad $R = mu/\hbar$ as before, we define S up to an additive function of t by

$$\text{grad } S = \frac{m}{\hbar}\left(v + \frac{e}{mc}A\right),$$

and let

$$\psi = e^{R+iS}.$$

Then ψ satisfies the Schrödinger equation

$$(11) \qquad \frac{\partial\psi}{\partial t} = -\frac{i}{2m\hbar}\left(-i\hbar\nabla - \frac{e}{c}A\right)^2\psi - \frac{ie}{\hbar}\varphi\psi + i\alpha(t)\psi,$$

where as before $\alpha(t)$ is real and can be made 0 by a suitable choice of S.

To prove (11), we perform the differentiations and divide by ψ, obtaining

$$\frac{\partial R}{\partial t} + i\frac{\partial S}{\partial t} = i\frac{\hbar}{2m}\left(\Delta R + i\Delta S + [\text{grad }(R + iS)]^2\right)$$

$$+ \frac{e}{mc}A\cdot\text{grad}(R + iS) + \frac{1}{2}\frac{e}{mc}\text{div } A - \frac{ie^2}{2m\hbar c^2}A^2 - \frac{ie}{\hbar}\varphi + i\alpha(t).$$

This is equivalent to the pair of equations we obtain by taking gradients and separating real and imaginary parts. For the real part we find

$$\frac{m}{\hbar}\frac{\partial u}{\partial t} = -\frac{\hbar}{2m}\frac{m}{\hbar}\,\mathrm{grad\ div}\,(v+\frac{e}{mc}A)$$

$$-\frac{\hbar}{2m}\,\mathrm{grad}[2\,\frac{m}{\hbar}\,u\cdot\frac{m}{\hbar}(v+\frac{e}{mc}A)] + \mathrm{grad}[\frac{e}{mc}A\cdot\frac{m}{\hbar}(v+\frac{e}{mc}A)]$$

$$+\frac{1}{2}\frac{e}{mc}\,\mathrm{grad\ div}\,A,$$

which after simplification is the same as the first equation in (2).

For the imaginary part we find

$$\frac{m}{\hbar}(\frac{\partial v}{\partial t}+\frac{e}{mc}\frac{\partial A}{\partial t}) =$$

$$\frac{\hbar}{2m}[\frac{m}{\hbar}\,\mathrm{grad\ div}\,u + \mathrm{grad}\,\frac{m}{\hbar}u\cdot\frac{m}{\hbar}u - \mathrm{grad}\,\frac{m}{\hbar}(v+\frac{e}{mc}A)\cdot\frac{m}{\hbar}(v+\frac{e}{mc}A)]$$

$$+\,\mathrm{grad}\,\frac{e}{mc}A\cdot(v+\frac{e}{mc}A) - \frac{e^2}{2m\hbar c}A^2 - \frac{e}{\hbar}\varphi.$$

Using (9) and simplifying, we obtain

$$\frac{\partial v}{\partial t} = \frac{e}{m}E + \frac{\hbar}{2m}\,\mathrm{grad\ div}\,u + \frac{1}{2}\,\mathrm{grad}\,u^2 - \frac{1}{2}\,\mathrm{grad}\,v^2.$$

Next we use the easily verified vector identity

$$\frac{1}{2}\,\mathrm{grad}\,v^2 = v\times\mathrm{curl}\,v + v\cdot\nabla v$$

and the fact that u is a gradient to rewrite this as

$$\frac{\partial v}{\partial t} = \frac{e}{m}E - v\times\mathrm{curl}\,v + u\cdot\nabla u - v\cdot\nabla v + \frac{\hbar}{2m}\Delta u. \tag{11}$$

But curl $(v+eA/mc)=0$, since the generalized momentum $mv+eA/c$ is by assumption a gradient, so that, by (8), we may substitute eH/mc for curl v, so that (11) is equivalent to the second equation in (2) with $F=ma$.

References

There is a long history of attempts to construct alternative theories to quantum mechanics in which classical notions such as particle trajectories continue to have meaning.

[46]. L. de Broglie, Étude critique des bases de l'interprétation actuelle de la mécanique ondulatoire, Gauthiers-Villars, Paris, 1963.

[47]. D. Bohm, A suggested interpretation of the quantum theory in terms of "hidden" variables, Physical Review 85(1952), 166-179.

[48]. D. Bohm and J. P. Vigier, Model of the causal interpretation of quantum theory in terms of a fluid with irregular fluctuations, Physical Review 96(1954), 208-216.

The theory which we have described in this section is (in a somewhat different form) due to Fényes.

[49]. Imre Fényes, Eine wahrscheinlichkeitstheoretische Begründung und Interpretation der Quantenmechanik, Zeitschrift für Physik 132 (1952), 81-106.

[50]. W. Weizel, Ableitung der Quantentheorie aus einem klassischen, kausal determinierten Model, Zeitschrift für Physik 134(1953), 264-285; Part II, 135(1953), 270-273; Part III, 136(1954), 582-604.

[51]. E. Nelson, Derivation of the Schrödinger equation from Newtonian mechanics, Physical Review 150(1966).

§16. Comparison with quantum mechanics

We now have two quite different probabilistic interpretations of the Schrödinger equation: the quantum mechanical interpretation of Born and the stochastic mechanical interpretation of Fényes. Which interpretation is correct?

It is a triviality that all measurements are reducible to position measurements, since the outcome of any experiment may be described in terms of the approximate position of macroscopic objects. Let us suppose that we observe the outcome of an experiment by measuring the exact position at a given time of all the particles involved in the experiment, including those constituting the apparatus. This is a more complete observation than is possible in practice, and if the quantum and stochastic theories cannot be distinguished in this way then they cannot be distinguished by an actual experiment. However, for such an ideal experiment stochastic and quantum mechanics give the same probability density $|\psi|^2$ for the position of the particles at the given time.

The physical interpretation of stochastic mechanics is very different from that of quantum mechanics. Consider, to be specific, a hydrogen atom in the ground state. Let us use Coulomb units; i.e. we set $m=e^2=\hbar=1$, where m is the reduced mass of the electron and e is its charge. The potential is $V=-1/r$, where $r=|x|$ is the distance to the origin, and the ground state wave function is

$$\psi = \frac{1}{\sqrt{\pi}} e^{-r}.$$

In quantum mechanics, ψ is a complete description of the state of the system. According to stochastic mechanics, the electron performs a Markoff process with

$$dx(t) = -\frac{x(t)}{|x(t)|}dt + dw(t),$$

where w is the Wiener process with diffusion coefficient $\frac{1}{2}$. (The gradient of $-r$ is $-x/r$.) The invariant probability density is $|\psi|^2$. The electron moves in a very complicated random trajectory, looking locally like a trajectory of the Wiener process, with a constant tendency to go towards the origin, no matter which direction is taken for time. The similarity to ordinary diffusion in this case is striking.

How can such a classical picture of atomic processes yield the same predictions as quantum mechanics? In quantum mechanics the positions of the electron at different times are non-commuting observables and so (by Theorem 14.1) cannot in general be expressed as random variables. Yet we have a theory in which the positions are random variables.

To illustrate how conflict with the predictions of quantum mechanics is avoided, let us consider the even simpler case of a free particle. Again we set $m=\hbar=1$. The wave function at time 0, ψ_0, determines the Markoff process. To be concrete let us take a case in which the computations are easy, by letting ψ_0 be a normalization constant times $\exp(-|x|^2/2a)$, where $a>0$. Then ψ_t is a normalization constant times

$$\exp\left(-\frac{|x|^2}{2(a+it)}\right) = \exp\left(-\frac{|x|^2(a-it)}{a^2+t^2}\right).$$

Therefore (by §15)

$$u = -\frac{a}{a^2+t^2}x,$$

$$v = \frac{t}{a^2+t^2}x,$$

$$b = \frac{t-a}{a^2+t^2}x.$$

Thus the particle performs the Gaussian Markoff process

$$dx(t) = \frac{t-a}{a^2+t^2}xdt + dw(t),$$

where w is the Wiener process with diffusion coefficient $\frac{1}{2}$.

Now let X(t) be the quantum mechanical position operator at time t (Heisenberg picture). That is,

$$X(t) = e^{\frac{1}{2}it\Delta}X_0e^{-\frac{1}{2}it\Delta},$$

where X_0 is the operator of multiplication by x. For each t the probability, according to quantum mechanics, that if a measurement of X(t) is made the particle will be found to lie in a given region B of $\mathbb{R}^3$ is just the integral over B of $|\psi_t|^2$ (where ψ_t is the wave function at time t in the Schrödinger picture). But $|\psi_t|^2$ is the probability density of x(t) in the above Markoff process, so this integral is equal to the probability that x(t) lies in B.

We know that the X(t) for varying t cannot simultaneously be represented as random variables. In fact, since the particle is free,

(1) $$X\left(\frac{t_1+t_2}{2}\right) = \frac{X(t_1)+X(t_2)}{2}$$

for all t_1, t_2, and the corresponding relation is certainly not valid for the random variables x(t). Thus the mathematical structures of the quantum and stochastic theories are incompatible. However, there is no contradiction in measurable predictions of the two theories. In fact, if one attempted to verify the quantum mechanical relation (1) by measuring

$$x(\frac{t_1 + t_2}{2}), \quad x(t_1), \quad x(t_2),$$

then, by the uncertainty principle, the act of measurement would produce a deviation from the linear relation (1) of the same order of magnitude as that which is already present in the stochastic theory of the trajectories. Although the operators on the two sides of (1) are the same operator, it is devoid of operational meaning to say that the position of the particle at time $(t_1+t_2)/2$ is the average of its positions at times t_1 and t_2.

The stochastic theory is conceptually simpler than the quantum theory. For instance, paradoxes related to the "reduction of the wave packet" (see §14) are no longer present, since in the stochastic theory the wave function is no longer a complete description of the state. In the quantum theory of measurement the consciousness of the observer (i.e., my consciousness) plays the rôle of a deus ex machina to introduce randomness, since without it the quantum theory is completely deterministic. The stochastic theory is inherently indeterministic.

The stochastic theory raises a number of new mathematical questions concerning Markoff processes. From a physical point of view, the theory is quite vulnerable. We have ignored a vast area of quantum mechanics - questions concerning spin, bosons and fermions, radiation, and relativistic covariance. Either the stochastic theory is a curious accident or it will generalize to these other areas, in which case it may be useful.

The agreement between the predictions of quantum mechanics and stochastic mechanics holds only for a limited class of forces. The Hamiltonians we considered (§15) involved at most the first power of the velocity in the interaction part.. Quantum mechanics can treat much more general Hamiltonians, for which there is no stochastic theory. On the other hand, the basic equations of the stochastic theory (Eq.(2) of §15 with F=ma) can still be formulated for forces which are not derivable from a potential. In this case we can no longer require that v be a gradient and no longer have the Schrödinger equation. In fact, quantum mechanics is incapable of describing such forces. If there were a fundamental force in nature with a higher order dependence on velocity or not derivable from a potential, at most one of the two theories could be physically correct.

Comparing stochastic mechanics (which is classical in its descriptions) and quantum mechanics (which is based on the principle of complementarity), one is tempted to say that they are, in the sense of Bohr, complementary aspects of the same reality. I prefer the viewpoint that Schrödinger [30, §14] expressed in 1926:

"... It has even been doubted whether what goes on in the atom could ever be described within the scheme of space and time. From the philosophical standpoint, I would consider a conclusive decision in this sense as equivalent to a complete surrender. For we cannot really alter our manner of thinking in space and time, and what we cannot comprehend within it we cannot understand at all. There _are_ such things - but I do not believe that atomic structure is one of them."

Milton Keynes UK
Ingram Content Group UK Ltd.
UKHW020753150824
446937UK00006B/204

9 780691 079509